THE SPECIFIC TREATMENT OF
VIRUS DISEASES

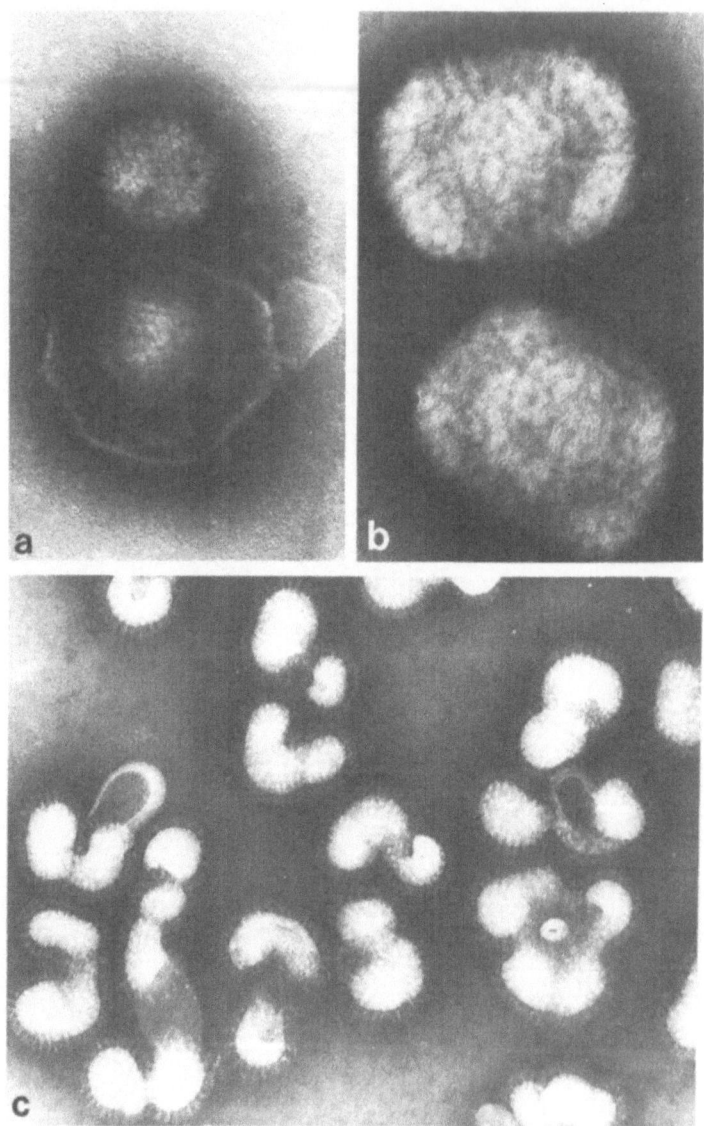

Viruses susceptible to chemotherapy (magnification × 180 000). (a) Herpes: the upper particle has lost its capsule and shows structural detail. (b) Vaccinia: two brick-shaped particles showing surface detail. (c) Influenza A: the projections on the surface are haemagglutinin and neuraminidase. [Photographs by courtesy of Dr. J. D. Almeida].

THE SPECIFIC TREATMENT OF VIRUS DISEASES

by

D. J. Bauer

M.A., Ph.D., M.B., B.Chir.
M.R.C.P., F.R.C.Path.

Wellcome Research Laboratories
Beckenham, Kent, England

MTP

Published by

MTP Press Limited
St. Leonard's House
St. Leonardgate
LANCASTER

Copyright © 1977 D. J. Bauer
Softcover reprint of the hardcover 1st edition 1977

No part of this book may be
reproduced in any form without
permission from the publishers,
except for the quotation of
brief passages for the purpose
of review.

ISBN 978-94-011-7923-2   ISBN 978-94-011-7921-8 (eBook)
DOI 10.1007/ 978-94-011-7921-8

Typeset by The Lancashire Typesetting Co. Ltd,
Bolton and printed by
R. & R. Clark Ltd, Edinburgh

Contents

Foreword

In comparison with antibacterial chemotherapy the clinical use of specific antiviral drugs is a fairly recent development. By the 1950s a number had been discovered, but their clinical application was delayed until 1962, when methisazone was used for the treatment of infective complications of smallpox vaccination and idoxuridine in the treatment of herpetic keratitis.

In retrospect the reasons for this slowness in development is now evident. It was due in large part to the prevailing view held at the time that virus infections could never be treated with specific agents, since the multiplication of virus is so closely integrated with the metabolic processes of the host cell that specificity of action seemed impossible of achievement. In contrast, bacteria were independent entities with metabolic systems differing widely from those of the mammalian host, which afforded points of attack by specific agents which had no action upon mammalian systems. Although the mode of action of the sulphonamides and penicillin was not known for some time, they afforded perfect examples of the success of this mode of attack.

It is now known that a number of viruses contain enzymes which are distinct from those of the host cell, and others carry the information for synthesizing virus-specific enzymes in their genetic code, thus affording points for specific attack by antiviral agents. Also, specificity is not always as essential as it may seem. For example, the synthetic nucleosides used in the treatment of infections with members of the herpesvirus group inhibit steps in the pathways of DNA synthesis of both virus and host cell, but inhibition of the multiplication of the virus can nevertheless be achieved by concentrations below those which affect the host cell. In this case specificity is lacking, but advantage can be taken of a favourable therapeutic ratio.

The way in which the subject of antiviral chemotherapy has developed inevitably means that it will be of less concern to general practitioners than to consultants in ophthalmology, dermatology and venereology. Nevertheless, an attempt has been made to present the subject on two different levels.

Firstly, as a text-book on antiviral therapeutics for medical students, who may abstract its salient points as they would a text-book of general medicine and pass over the details. For this purpose a general reading list is appended to each chapter. Secondly, as a reference work for postgraduate students preparing for the examinations of the Royal Colleges who may need to study the subject in greater detail, and also for consultants who wish to have a comprehensive assessment of the relevant literature in a single volume. For such readers the text is provided with references to the original work which are arranged alphabetically in a bibliography at the end of the work. The illustrations have been restricted in number and show clinical material where possible. Illustrations of disease conditions in untreated subjects have not been provided, since these are available in the standard texts which are quoted as general reading. Descriptions of treatment are confined to the use of antiviral agents, and it should be understood that additional measures such as antibiotics, steroids and intensive care may be necessary, and information of this kind may be obtained by consulting the relevant references.

The selection of compounds included in this book has been limited to those which are actually in clinical use as antiviral agents, and they are all to be found in the 25th edition of Martindale's Extra Pharmacopoeia. For this reason no attempt has been made to delve into the controversy as to the value of ascorbic acid in the prophylaxis of colds, since although it is extensively used for this purpose by the general public it is devoid of specific antiviral activity.

I am deeply indebted to my secretary, Mrs. B. P. Moore, for her constant encouragement and her assistance in preparing the manuscript for publication. My thanks are also due to Mr. J. W. T. Selway for assistance in preparing the illustrations, to Dr. K. Apostolov, Royal Postgraduate Medical School for supplying the photographs for Figure 5.1, and to Dr. J. D. Almeida, who kindly supplied the electron micrographs of the frontispiece.

D. J. Bauer

Chapter 1

An introduction to the viruses

Viruses are obligate intracellular parasites of man, other animals, plants and bacteria. They can only multiply within cells, and thus differ from bacteria, which can multiply in tissues in an extracellular position, and also in artificial culture media in the laboratory.

It has always been recognized that viruses are distinct from the bacteria, but originally other groups of infective agents were included among the viruses which are now known to be organisms of a different and more complex nature. These included the Chlamydiae, organisms causing diseases such as lymphogranuloma venereum and trachoma, and the Rickettsiae, the causative agents of typhus and related infections. The specific treatment of infections caused by these two groups of agents will not be considered in this work, which is restricted to the chemotherapy of infections caused by the true viruses. The Chlamydiae and Rickettsiae are complete organisms, with cell walls, which possess both types of nucleic acid, enzyme systems which function within their cytoplasm, and a number of cell organelles. The viruses, on the other hand, possess either DNA or RNA as their genetic material, but not both. In some cases they contain enzymes, but these exert their functions within the host cell after the virus particle has undergone a process of dissolution during the early stage of infection. Except in the case of the arenaviruses, a group which contains the virus of Lassa fever, the virus particles do not contain organelles. A virus is thus much more rudimentary in structure, and relies upon the host cell to provide the systems for synthesizing its components which it does not possess itself. This is the underlying reason for its obligate intracellular parasitism and inability to multiply in any other situation.

CLASSIFICATION OF THE VIRUSES

The scope of this work is limited to those viruses which cause disease in man, but in order to view them in proper perspective it is desirable to consider the place which they occupy in the classification of the animal viruses as a whole. The viruses fall sharply into two classes: those which utilize RNA as their genetic material, and those which use DNA. As stated earlier, viruses never contain both types of nucleic acid.

Further subdivision is based on structural features, such as symmetry and properties of the component parts. The complete virus particle is known as the virion. It has an outer protein coat known as the capsid. This may be associated with the virus nucleoprotein, in which case it is referred to as the nucleocapsid, or the nucleoprotein may be present as a central core, in which case the capsid may consist of an assemblage of repeating units known as capsomeres. The virion may have an outer envelope derived from the nuclear membrane or plasma membrane of the host cell.

The virion is usually symmetrical, and the symmetry may be helical or of the type known as cubic. These features are used in the construction of the presently accepted system of classification. The system used for the DNA viruses is shown in Table 1.1, which is based on a table given by Melnick (1974). Except for the parvoviruses all groups are of importance in human medicine.

A similar classification for the RNA viruses is given in Table 1.2. All groups contain viruses of clinical importance except the oncorna and coronavirus groups. Respiratory syncytial virus, an important cause of upper respiratory infection, occupies an intermediate position between the orthomyxovirus and paramyxovirus groups, and together with a virus causing pneumonia in mice may constitute an independent group of metamyxoviruses.

VIRUS MULTIPLICATION

The first stage in the multiplication of a virus involves the entry of the virion into the host cell. It first becomes attached to the plasma membrane of the host cell by electrostatic forces. In some cases the membrane contains receptor sites which are specific for the virus. The virion then passes into the interior of the cell, either by fusion of its outer coat with the plasma membrane, or by engulfment in a phagocytic vacuole. If fusion occurs the nucleocapsid may pass into the cell, but if the virion is taken into the cell by phagocytosis the outer membrane is removed by cell enzymes, a process known as uncoating. When this has taken place infectious virus is no longer present in the cell. This stage of the growth cycle is known as the eclipse phase.

When the genetic material of the virion is exposed it undergoes two processes. It must be replicated in order to provide sufficient genetic material for assembly in the progeny virions, and the genetic message which it carries must be transcribed and translated so that the polypeptides of the virus may be synthesized on the host cell ribosomes. These processes are complicated and will be described here only in outline.

Table 1.1 Classification of DNA viruses of animals and man (after Melnick, 1974)

	Parvovirus	Papovavirus*	Adenovirus	Herpesvirus	Poxvirus
Symmetry	Cubic				Complex
Envelope	None			Present	Complex
Site of assembly of capsid	Nucleus			Nucleus	Cytoplasm
Site of envelopment				Nuclear membrane	
Number of capsomeres	32	72	252	162	
Diameter of virion (nm)	18–22	45–53	70–90	100	230 × 300
Group	Parvovirus	PAPOVAVIRUS*	ADENOVIRUS*	HERPESVIRUS*	POXVIRUS*
Examples of infections in man		Wart virus	Adenovirus	Herpes Varicella-zoster Cytomegalovirus	Vaccinia Smallpox

* Where the group name is in capitals it has been designated as a family

Table 1.2   Classification of RNA viruses of animals and man (after Melnick, 1974)

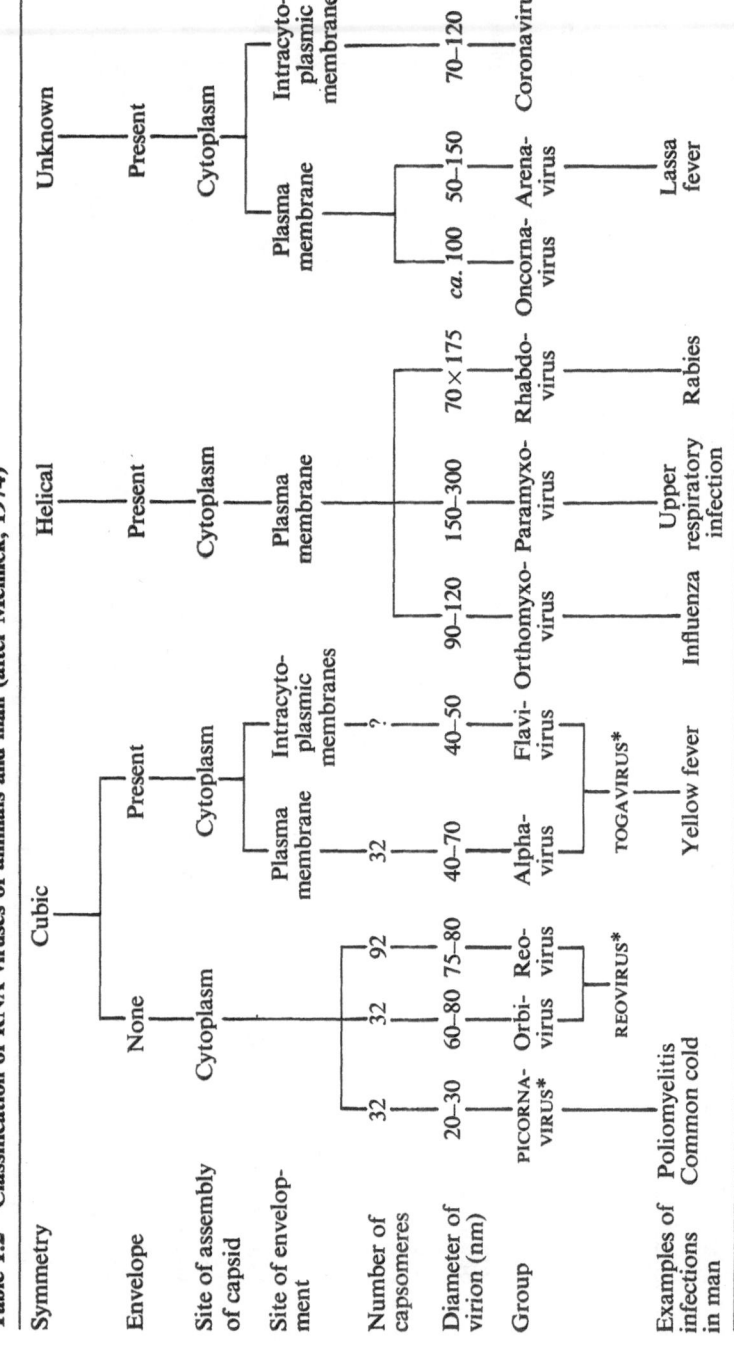

Symmetry	Cubic			Cubic		Helical			Unknown		
Envelope	None			Present		Present			Present		
Site of assembly of capsid	Cytoplasm			Cytoplasm		Cytoplasm			Cytoplasm		
Site of envelopment				Plasma membrane	Intracyto-plasmic membranes	Plasma membrane			Plasma membrane		Intracyto-plasmic membranes
Number of capsomeres	32	32	92	32	?						
Diameter of virion (nm)	20–30	60–80	75–80	40–70	40–50	90–120	150–300	70×175	ca. 100	50–150	70–120
Group	PICORNA-VIRUS*	Orbi-virus	Reo-virus	Alpha-virus	Flavi-virus	Orthomyxo-virus	Paramyxo-virus	Rhabdo-virus	Oncorna-virus	Arena-virus	Coronavirus
		REOVIRUS*		TOGAVIRUS*							
Examples of infections in man	Poliomyelitis Common cold			Yellow fever		Influenza	Upper respiratory infection	Rabies		Lassa fever	

* Where the group name is in capitals it has been designated as a family

The DNA and RNA viruses differ in the methods employed for replicating their genetic material. In the former the DNA of the virion is replicated by a DNA polymerase, either present in the host cell or in some cases coded for by the virus and synthesized on the host cell ribosomes. RNA viruses can replicate RNA directly from their RNA genome by using it as a template. This process forms complementary strands, which act as templates for the synthesis of RNA in which the genetic code is restored for the purpose of incorporation into the progeny virions.

In the process of transcription RNA strands are generated which carry the equivalent of the genetic code and act as messenger RNA for the translation of virus proteins on the host cell ribosomes. The RNA genome of the virion may act directly as a messenger RNA and be translated into a very large polypeptide which is subsequently broken down by enzymes into shorter units. In some viruses, however, the RNA of the virion has an arrangement of bases which is complementary to the required code. It must therefore be transcribed by an RNA polymerase in order to generate messenger RNA with the required sequence of bases. The RNA of the genome is in some cases double-stranded, and one strand is transcribed into messenger RNA by a transcriptase carried in the virion.

Translation from messenger RNA follows a similar course in both DNA and RNA viruses, giving rise to enzymes required for virus multiplication and structural polypeptides required for making the new generation of virions.

When the components of the virus have been synthesized they are assembled into virions which are then released from the cell. The details of this process, together with the general features of the growth cycle, will be described separately for those groups of viruses which are of importance in chemotherapy.

Herpesviruses

The stages in the multiplication of the herpesviruses have been critically reviewed by Watson (1973). The virus particles enter the cell by fusion to the plasma membrane, or by engulfment in pinocytotic vesicles.

The early events in the multiplication cycle of the herpesviruses take place in the nucleus, with the formation of strands of RNA coded for by the virus. These pass into the cytoplasm and break down into shorter lengths which act as messenger RNA for the synthesis of capsid proteins, which then pass back into the nucleus, where the virus DNA has been replicating meanwhile. The various components then become assembled into capsids, which are spherical bodies bounded by a membrane and usually containing a dense core, which is presumably nucleoprotein. At this stage the particles are still inside the nucleus, but particles consisting of a core surrounded by two membranes now begin to appear in the cytoplasm. These are the mature virions, and it is considered that the intranuclear capsids enter the cytoplasm by passing through the nuclear membrane, acquiring a second coating in the process. The virions are liberated from the cell by budding through the plasma membrane, or by lysis of the cell. The length of the growth cycle is about 15 hours. The eclipse phase lasts for 8 hours, and infectious virions appear 2 hours later and reach peak titres 15 hours after the onset of infection.

Poxviruses

The member of the poxvirus group which has been most studied is vaccinia. The stages in its multiplication cycle have been reviewed by Fenner *et al.* (1974). The virion enters the cell by pinocytosis, and while it is in the pinocytotic vesicle the outer coats are removed by host cell enzymes leaving the central core which contains the virus DNA. The limiting membrane of the vacuole disappears and the core enters the cytoplasm. The virion contains a DNA-dependent RNA polymerase which synthesizes virus-specific messenger RNA, and it is thought that this codes for an enzyme which disrupts the core and liberates the virus DNA. The virus also codes for a DNA polymerase, by means of which the progeny DNA is synthesized. This takes place in localized areas in the cytoplasm. The progeny DNA is now transcribed, and the messenger RNA thus formed is translated on the host cell ribosomes to form structural polypeptides. These aggregate to form capsules, beginning at the margins of the areas in the cytoplasm where virus DNA is being synthesized, and gradually extending to surround a mass of DNA and finally enclose it. The material inside differentiates into a core and two lateral bodies and assembly of the progeny virions is then complete. Infective virus appears after 4 hours and by 20 hours or so the cycle of growth is complete. Liberation from the cell takes place as the result of lysis.

Orthomyxoviruses

The member of the orthomyxovirus group which is of importance in chemotherapy is influenza A. The virion contains seven strands of ribonucleoprotein and is bounded by a capsule which contains a haemagglutinin and the enzyme neuraminidase, which occur in the form of spikes projecting from the surface.

The existing state of knowledge of the replication cycle of influenza virus has been reviewed by Fenner *et al.* (1974). Although it is an RNA virus its replication involves the nucleus at some stage, since influenza virus will not multiply in cells which have been enucleated with cytochalasin B, or in which the nuclear DNA has been damaged by exposure to ultraviolet light. The virion attaches to the cell by means of the haemagglutinin, which reacts with a specific receptor in the plasma membrane. The mechanism of penetration has not been established beyond doubt. There is evidence that the envelope of the virion fuses with the plasma membrane and also that the virions are taken up into pinocytotic vesicles. The strands of ribonucleoprotein contain a RNA polymerase which transcribes a messenger RNA which is complementary in its base sequence to the virus RNA. The proteins of the virus are synthesized by translation of the messenger RNA on the cytoplasmic ribosomes, but after synthesis they migrate to other parts of the cell. The protein forming the nucleocapsid migrates to the nucleus, and the haemagglutinin and protein of the virion membrane accumulate on the smooth endoplasmic reticulum. A non-structural protein is also formed, which may have a function in regulating transcription of RNA. It passes into the nucleolus. During the later stages of the growth cycle the RNA of the virus begins to replicate, so that virion RNA is formed at the expense of the complementary RNA synthesized during the early stages of infection. Assembly of the virions and their release from the cell take place as parts of a single process. Areas of plasma membrane develop visible projections, representing units of neuraminidase and haemagglutinin, and beneath them accumulations of ribonucleoprotein and a layer of membrane protein appear. The whole complex buds from the cell surface and becomes nipped off to form a complete virion. It is thought that the neuraminidase plays some part in enabling the virion to be released from the cell.

THE COURSE OF VIRUS INFECTIONS

After liberation from the cell at the end of the first cycle of infection the progeny virions may infect the cells in immediate contact, or may gain entry into the blood stream and give rise to infection in other parts of the body. The type of illness which ensues depends upon which process predominates. It may be acute, recurrent or chronic.

Acute infections

The acute virus infections can be divided into two groups: those in which the infection affects a single organ or tissue, and those in which an initial period of multiplication in some site, which may be entirely asymptomatic, is followed by a period of dissemination in which the virus multiplies in another tissue to give the characteristic features of the disease.

Typical examples of the first group are the common cold and influenza. The initial infection takes place in the nasal or bronchial mucosa, and the infection spreads laterally to involve a considerable part of the cell layer.

Chickenpox and smallpox are characteristic examples of the second group. The initial site of multiplication is uncertain, possibly the reticulo-endothelial system or the respiratory tract. The virus undergoes several cycles of multiplication, until so much virus is produced that it spills over into the blood stream to produce a viraemia. The virus then undergoes further cycles of multiplication in the skin, producing the lesions characteristic of the disease. The extent of the skin involvement depends upon the number of virus particles liberated into the blood stream during the phase of viraemia. In both chickenpox and smallpox less than a dozen lesions may be present and the disease may escape detection. In other cases the eruption may be very extensive and associated with severe general illness.

Acute virus infections bring about an immune response which usually terminates the infection, with subsequent recovery. In cases of smallpox with a confluent eruption the immune mechanisms may be insufficient to arrest the infection and the death of the patient ensues.

The immune response resulting from an acute infection may often persist for many years. As a result a second attack may never occur, or only very rarely. However, second and further attacks occur quite commonly in a number of virus infections. There are two reasons for this. Firstly, the virus may not be completely eliminated from the body, but may persist in some site where it is not exposed to antibody. As a result of some stimulus it may become reactivated and give rise to a further attack of the disease. This behaviour is characteristic of viruses of the herpes group, and will be described in more detail in the chapters on the specific treatment of herpes, varicella-zoster and cytomegalic inclusion disease.

Secondly, further attacks may be due to infection with a virus of a serotype different from that of the initial infection. This situation is typical of the respiratory infections. There are at least 120 serotypes of rhinovirus which have no cross-immunity. Infection with one serotype leads to lasting immunity, but there are so many serotypes that fresh attacks of the common cold are a constant occurrence. Influenza virus undergoes a radical change of antigenicity every 10 years or so, a process known as antigenic shift. The general population has little or no immunity to the new type and an epidemic thus occurs. Immunity to the new type persists, but it undergoes further minor changes known as antigenic drift, which reduce the effectiveness of the existing antibody, thus allowing further attacks to occur.

Chronic virus infections

Certain virus infections are characterized by a chronic course which may last for weeks or months. Typical examples are warts and molluscum contagiosum. Here the infection does not cause lysis of the cell, so that the virus is not exposed to antibody. The cells are stimulated to proliferate and give rise to benign tumours, which increase in diameter by direct infection of cells lying on either side.

In patients in whom the ability to form antibody is congenitally absent or reduced, or suppressed by treatment with immunosuppressant drugs, an infection which is normally acute may pursue a chronic course. A typical but rare example which is invariably fatal in the absence of treatment is vaccinia gangrenosa. In this condition the lesion of primary smallpox vaccination fails to heal, and continues to enlarge and give rise to metastatic lesions over a period of several months until the extent of tissue destruction is incompatible with survival. In transplant patients under immunosuppression the skin lesions of herpes may pursue a very prolonged course.

The virus infections considered so far have all been of the productive type, in which the growth cycle terminates in the formation of infective virions which proceed to spread the infection. In some conditions, however, infective virions are formed less frequently or not at all. Cell damage then progresses much less rapidly and a chronic infection is set up. Examples of this situation are seen in subacute sclerosing panencephalitis and progressive multifocal leukoencephalopathy. In the former condition electron microsocpy reveals microtubules resembling those of the myxoviruses in the nuclei of neurones, and strains of virus resembling measles have been isolated from brain biopsy material in tissue culture. In progressive multifocal leukoencephalopathy particles resembling those of a papovavirus can be found.

Transforming infections

The main feature of warts and molluscum contagiosum is a benign cell proliferation with production of infective virions. With some viruses this situation is carried a stage further; the infection induces a malignant transformation in the cells, which then proliferate in an uncontrolled manner, and the production of infective virions is reduced or completely abolished. A number of RNA viruses of animals behave in this manner, such as Rous sarcoma virus and various murine leukaemia viruses, and it is a theoretical possibility that a number of human malignant conditions result from infection with a virus of this type. Similar examples occur among the DNA viruses as well. Certain serotypes of human adenovirus will transform cells *in vitro* and induce malignant tumours in hamsters. Marek's disease of chickens is a malignant lymphomatosis due to infection with a virus of the herpes group.

Cell transformation by DNA viruses is of importance in human medicine also. There is some evidence that infection with type 2 herpes virus is linked with the development of cervical carcinoma. Epstein–Barr virus is a virus belonging to the herpes group which is present in a latent form in the cells of Burkitt's lymphoma. The part it plays in the causation of this tumour has not been established beyond doubt, but the extensive laboratory work carried out with the virus led to the chance finding that it is the causative agent of infectious mononucleosis. Lymphocytes infected with the virus undergo transformation into cells of lymphoblast type which will grow indefinitely in culture. The disease therefore pursues a prolonged course and it is not clear by what means it is brought to an end. Compounds are available which will inhibit the multiplication of viruses belonging to the herpes group, as will be described later, and it is therefore possible that infectious mononucleosis might respond to specific chemotherapy.

Slow virus infections

There are certain conditions with a protracted course which have features which set them apart from the chronic virus infections described already. Both types of virus hepatitis are characterized by a very long incubation period. The viruses responsible have not been isolated and adapted to growth in tissue culture, so that little is known of their properties, but it is likely that the growth cycle is exceptionally long.

There is considerable evidence that multiple sclerosis is a slow virus infection which gives rise to an auto-immune response. Both measles (Ter Meulen et al., 1972 a) and parainfluenza type 1 (Gudnadóttir et al., 1964) have been isolated from the plaques, but evidence supporting an aetiological role for these viruses has been difficult to obtain. A more promising candidate is a virus isolated by Carp et al. (1972). Although it is apparently present in large amounts it has not yet been possible to characterize it. Specific treatment of multiple sclerosis with antiviral agents is thus becoming a theoretical possibility.

Subacute myelo-optic neuropathy is a degenerative disease of the central nervous system which was first reported in Japan. There is evidence that it may also occur in Australia, Great Britain and the United States. It was initially attributed to a toxic side-effect of clioquinol, but it has now been shown to be due to infection with a virus belonging to the herpes group (Inoue, 1975). It is characterized by symmetrical degeneration of the posterior and lateral tracts of the spinal cord, peripheral nerves and sometimes the optic nerve, leading to sensory disturbances in the legs, muscular weakness, bilateral impairment of visual acuity, and occasionally psychosis. The virus is thought to be a variant of avian infectious laryngotracheitis virus, a member of the herpes virus group which affects poultry. It would seem reasonable to treat the disease with the standard drugs used for treating herpes and varicella-zoster infections, but there are no reports so far of this having been done.

Finally, there is a group of chronic degenerative conditions of the central nervous system caused by infective agents of virus nature which have not yet been visualized in the electron microscope. They are thought to be short strands of RNA unassociated with structural protein. The most studied example is scrapie, a chronic disease of sheep. Kuru and Creutzfeldt–Jakob disease are caused by agents of this type. Both have been transmitted to primates, and the two diseases are now considered to be identical. Other chronic neurological conditions which may possibly belong to the same group are Alzheimer–Pick disease, amyotrophic lateral sclerosis and paralysis agitans, and it is not beyond the bounds of possibility that these intractable conditions may eventually respond to antiviral chemotherapy.

GENERAL READING

Andrewes, C. H. and Pereira, H. G. (1972). *Viruses of Vertebrates.* 3rd edn. (London: Baillière Tindall).

Fenner, F., McAuslan, B. R., Mims, C. A., Sambrook, J. and White, D. O. (1974). *The Biology of Animal Viruses.* (London: Academic Press Inc.).

Chapter 2

Viruses in relation to chemotherapy

In the preceding chapter the features of virus infections were outlined in general terms with more particular regard to immunity and events at the cellular level, but from the standpoint of chemotherapy it is important to consider how these events control the clinical course of the disease process, since this will determine whether the disease is likely to be susceptible to specific treatment or not.

In what follows the various types of clinical course will be analysed without relevance to whether specific agents for treating the disease have already been found. The situation in antiviral chemotherapy at the present time resembles in many ways that of the chemotherapy of bacterial diseases in the early 1940s, when the sulphonamides and penicillin were available as effective agents but the wide range of antibiotics now at the disposal of the clinicians still awaited discovery.

Acute infections

Some of the commonest virus diseases are acute infections with a duration of illness so short that there would seem to be little time to institute specific treatment before recovery sets in. Typical examples are the common cold and influenza. However, the general public are very quick to diagnose the onset of these conditions, and there are many folk remedies which are regularly taken with the object of alleviating the symptoms of the common cold. A number of compounds which inhibit the multiplication of rhino-viruses in tissue culture have been brought to clinical trial but none has as yet proved effective. It will be shown later that the duration of illness in influenza can be significantly reduced by the administration of amantadine hydrochloride.

Other virus diseases which are classed as acute may have a more protracted course. From the chemotherapeutic aspect they may be considered as falling into two groups, according to whether there is a prodromal phase or not. Examples of the former are adenovirus infections, infectious mononucleosis, chickenpox and mumps. Zoster may also be included here, since the pain which may precede the eruption may attract attention. The onset of a typical adenovirus infection is gradual, with increasing malaise, fever, conjunctivitis and pharyngitis. The febrile period may last for a week or more, and the conjunctivitis may persist for longer still. Methisazone and trifluorothymidine will inhibit the multiplication of adenoviruses in tissue culture, but they have not proved effective in clinical use (Little *et al.*, 1968).

In infectious mononucleosis the illness may last for up to 3 weeks or more. As mentioned already, the condition is caused by a virus belonging to the herpes group, and it might well respond to treatment with the compounds in clinical use which inhibit the multiplication of herpes and varicella-zoster. The lesions of varicella appear in crops over a period of a week or so, making it possible to give specific antiviral treatment in cases where this is considered justified. The lesions of zoster pursue a protracted course of development, and adequate time is available for giving specific treatment.

A prodromal phase gives warning in advance of the disease which is going to develop, and affords time for the institution of specific treatment. It occurs in measles, smallpox and infective and serum hepatitis. The onset of measles is characterized by fever, coryza, conjunctivitis and an exanthem, which takes the form of red macules (Koplik's spots) on the buccal mucosa. Measles virus is inhibited by 6-azauridine; this has been used in malignant conditions but its use in the treatment of measles would not be considered justified. The prodromal phase in smallpox lasts for 2 or 3 days, and is characterized by the acute onset of fever, severe headache, back pain and prostration. It is easily recognized and there is usually a history of recent contact. Methisazone administered during the prodromal phase may arrest the infection or mitigate the further course in the exanthematic phase. Infective and serum hepatitis usually begin with a preicteric phase, with anorexia, nausea, vomiting and fever which may last up to 21 days.

Some acute virus infections may pursue a biphasic course. Poliomyelitis may begin with a minor illness of influenzal type lasting for about a week, followed by the major illness with paralysis after a symptom-free interval. The multiplication of poliomyelitis virus in tissue culture can be effectively inhibited by 2-(α-hydroxybenzyl)benzimidazole, but the virus becomes resistant after a few cycles of multiplication and the compound has therefore not been developed for clinical use. The success of vaccination in eradicating the disease has also meant that antiviral chemotherapy now has little part to play. A biphasic course also occurs in echovirus meningitis and tick-borne encephalitis.

Chronic virus infections

Chronic virus infections characteristically involve the skin and central nervous system. Warts, molluscum contagiosum and vaccinia have already been mentioned, and the chronic nature of the course affords ample time for instituting specific treatment. The viruses of the herpes group cause chronic infections of a special type, in which the initial acute phase leading to recovery and the development of humoral immunity is followed by a period of latency during which the virus persists in the central nervous system. Reactivation may take place at any time, with recurrence of the local lesions. An important example of this type is herpetic keratitis, and recurrences of this condition can be prevented by the prophylactic administration of idoxuridine.

An important use of antiviral chemotherapy is in prophylaxis. This differs from prophylaxis with vaccines in an important respect. Vaccines confer immunity against infection which is going to take place in the future. The immunity takes some time to develop, and if infection has already occurred in a person whose immunity is low or absent, little effect can be expected from vaccination. Thus, vaccination or revaccination after contact with smallpox cannot be relied upon to protect the contact absolutely although it may moderate the severity of the clinical illness. In contrast, an antiviral compound administered prophylactically will exert its inhibiting effect as soon as it is present in the tissues in sufficient concentration, and bring the infection to a halt before the onset of clinical illness. This use is clearly prophylaxis from the viewpoint of public health, since the intent is to prevent a future event, but it is actually the specific treatment of an infection which may already be in progress but which has not yet developed to the stage of clinical illness. Examples of this type of use are the prophylaxis of smallpox with methisazone and the prophylaxis of influenza with amantadine hydrochloride.

Antiviral compounds may also be used prophylactically to prevent recurrences of an infection which has entered a latent phase. An example of this is the use of eyedrops of idoxuridine and trifluorothymidine to prevent recurrences of herpetic keratitis.

Antiviral agents in comparison with vaccines and antisera

An important advantage which antiviral agents have over vaccines has already been mentioned, in that they can be used prophylactically against an infection which is already in being. They are also generally easier to manufacture and have fewer restrictions on their storage life. Effectiveness and freedom from unacceptable toxicity can be established once and for all and further tests on subsequent batches are unnecessary, except for the usual control of quality generally required for therapeutic substances.

With vaccines, on the other hand, each batch must be tested for efficacy and freedom from pathogens, so that a substantial fraction of each batch may have to be used up for control purposes.

The discovery of antiviral compounds with activity against a particular virus infection is a matter of chance, and it is as yet not possible to design a compound with the activity required. It is a relatively simple matter to do this with vaccines, by inactivating the virus or attenuating it until it produces immunity without symptoms of illness. The successes achieved by vaccines are impressive, and their use has brought about the virtual eradication of poliomyelitis, yellow fever and smallpox. Vaccines have been less successful in controlling the respiratory infections caused by viruses. Vaccination against the common cold is impracticable, since there are at least 120 serotypes of rhinovirus which are serologically unrelated, so that an effective vaccine would need to have a very large number of components. Influenza virus undergoes a major change (antigenic shift) every 10 years when a new strain emerges which is serologically unrelated to the strain which has hitherto prevailed, so that the existing vaccine is ineffective. In addition, the virus may undergo minor changes (antigenic drift) which make the vaccine less effective. The common cold and influenza are thus major targets for chemotherapy.

Vaccines are not of value in the treatment of virus infections except in the case of rabies, where vaccine is given after infection has occurred but before the onset of clinical illness. Immune sera may be used for passive immunization, as in the case of the administration of human gamma-globulin in persons exposed to infective hepatitis. They are not generally of value in treatment, except in the case of eczema vaccinatum and vaccinia gangrenosa, which may respond favourably to the administration of vaccinia hyperimmune gamma-globulin.

Possible sites of antiviral action

In order to give rise to infection the virus particle must gain entrance to the body and infect a susceptible cell. Before infection takes place the virus particle is lying free, on a mucous membrane or in the tissue space of the target organ. In such a position bacteria can multiply, produce toxins and cause symptoms, but the free-lying virus particle is completely inert, since multiplication can only take place after penetration of the cell has occurred. In this free-lying position the virus can be neutralized by circulating antibody, but cannot be attacked by chemotherapy, since compounds which could inactivate the virus particle without causing damage to the host are not available.

The first stage of infection is the attachment of the virus to the cell surface. In those cases in which the virus attaches to a specific receptor it is possible to block this process with therapeutic agents. Amantadine hydrochloride is thought to act in this way, by blocking the receptors for influenza virus. In the second stage of infection the virus particle passes through the plasma membrane of the cell into the cytoplasm, and its genetic material is then released. Amantadine hydrochloride is thought to inhibit this stage also.

The subsequent stages of synthesis of nucleic acid and proteins also afford sites of attack for antiviral agents. A number of synthetic nucleosides have been discovered which inhibit the synthesis of DNA in infections with herpesviruses, and protein synthesis by the poxviruses can be inhibited by methisazone.

Resistance in antiviral chemotherapy

Viruses can be made resistant to chemotherapy in the same way as bacteria. Thus, Buthala (1964) found that in tissue cultures infected with herpes virus and treated with high concentrations of idoxuridine the multiplication of the virus was inhibited for some time, but infectious virus was eventually produced, which remained resistant to idoxuridine through subsequent passages. Resistant strains also account for some cases of herpetic keratitis which fail to respond to treatment with idoxuridine. This does not pose any problems in treatment, however, since strains of herpes virus which are resistant to idoxuridine are sensitive to trifluorothymidine and cytarabine.

The type of cell in which the virus is multiplying may also be a factor in resistance to chemotherapy. The importance of cell type is evident in the case of the myxoviruses, which are unable to infect cells which lack the specific receptor on the plasma membrane. Influenza can thus only infect the cells of the respiratory mucosa. Other viruses can infect a much greater variety of cells, and in this case their sensitivity to antiviral agents may vary according to the type of cell in which they are growing. For instance, Appleyard et al. (1965) investigated the sensitivity of rabbitpox virus to isatin 3-thiosemicarbazone (the parent compound of methisazone) in HeLa cells in comparison with RK13 and L cells. They found that the yield of virus in HeLa cells was reduced by 2 log units in the presence of concentrations of 0·5–2 µg/ml, but that a concentration of 32 µg/ml was required to bring about a comparable degree of reduction in the other two cell lines. Cell-dependent resistance of this type may be the basis for the ineffectiveness of methisazone in the treatment of smallpox. The compound is effective in prophylaxis during the incubation period when the virus is probably growing in the cells of the reticulo-endothelial system, but it is possible that the virus is much less sensitive during the eruptive phase when it is growing in the cells of the dermis. It is also equally possible that the concentrations attained in the skin may be too low to exert an antiviral effect.

Resistance is not as yet of importance in antiviral chemotherapy. This is undoubtedly due to the fact that the major virus infections are not as yet widely treated with antiviral agents. Another important factor is that courses of treatment in antiviral chemotherapy are short. For both these reasons there is not much opportunity for the emergence of resistant strains.

Methods of antiviral chemotherapy

A knowledge of the methods used in testing antiviral activity will be of some assistance in understanding the chapters on antiviral drugs which follow, since they are of importance in determining the relative potencies, and also variations in chemotherapeutic sensitivities between strains and types of the same virus. Certain methods are also used for the determination of the concentrations of antiviral agents in tissue specimens and biological fluids.

The basic method for carrying out a test of antiviral activity consists of replicate titrations of a virus preparation carried out in parallel in normal tissue culture medium and medium containing the compound in a concentration which is below its toxic level. The titre of the virus preparation is determined by observation of the cytopathic effect, which is generally the destruction of the cell sheet which results from the virus infection, and is expressed as the ID50, that concentration of virus which suffices to produce cell destruction in 50% of the replicate cultures. If the compound has antiviral activity a reduction in ID50 will be observed.

Another method which is considerably more convenient for the detection of antiviral activity is the plaque inhibition test, which was developed independently by Herrmann (1961) and Rada *et al.* (1960). The method is based on the same principle as the disc test used for assessing the sensitivity of bacteria to antibiotics. A confluent monolayer of susceptible cells is grown on the bottom of a petri dish and infected with an amount of virus sufficient to produce a large number of plaques. After allowing sufficient time for adsorption the virus suspension is removed and the plate is covered with an agarose overlay. After solidification of the overlay a disc of filter paper which has been dipped in a solution of the test compound and dried is placed in the centre, and the plate is incubated further to allow plaques to develop. The compound will diffuse out from the disc, and if it possesses antiviral activity the underlying area of the cell sheet will be protected from infection. The disc will therefore be surrounded by an area free from plaques which extends out to the region where the concentration of the compound is insufficient to prevent the formation of plaques. The diameter of the plaque-free zone gives a rough measure of the extent of the activity of the compound. The method has been used by Herrmann *et al.* (1960) for the bioassay of solutions containing unknown amounts of antiviral agents.

A much more accurate measure of antiviral activity may be obtained by the method of plaque reduction. This is based on the fact that an antiviral agent incorporated in the agarose overlay will reduce the number of plaques in proportion to the logarithm of its concentration. Plates are set up as for the plaque inhibition test, except that the antiviral compound is incorporated in the overlay in a range of doubling dilutions. When the number of plaques which develop is expressed as a percentage of the number obtained in control plates from which the compound has been omitted and plotted against the logarithm of the concentration, a dose–response line is obtained. The situation of the dose–response line gives an absolute measure of antiviral activity, since it enables the concentration which gives 50% plaque reduction (ID50) to be calculated. Examples of such dose–response lines are given in Chapter 3, where they are used to establish the relative potencies of the standard compounds used in the treatment of herpesvirus infections.

GENERAL READING

Bauer, D. J. (1972). Introduction to antiviral chemotherapy. In: D. J. Bauer (ed.) *Chemotherapy of Virus Diseases*, vol. 1, 1–33. (Oxford: Pergamon Press).

Chapter 3

Antiviral agents—I

The antiviral nucleosides

INTRODUCTION

In the 1960s much work was devoted to the synthesis of analogues of the components of nucleic acids, with the object of developing antimetabolites with particular reference to the treatment of malignant conditions. A number of compounds emerged which were effective inhibitors of one or more stages in the pathways of DNA synthesis. In the treatment of malignant conditions the results obtained were generally disappointing, except in the case of cytarabine, which has found a lasting place in the treatment of leukaemia. These early studies were of great value, however, in that they provided an abundance of information on the clinical pharmacology and metabolism of these compounds, which is indispensable to physicians who now wish to use them in the treatment of virus infections.

It was recognized early that compounds which would inhibit the synthesis of DNA in tumour cells might well inhibit the multiplication of DNA viruses, such as herpes and vaccinia. This line of investigation soon proved very fruitful, and led to the introduction of idoxuridine and cytarabine into clinical practice for the treatment of infections caused by viruses of the herpes group. This particular field is still yielding compounds of value, and recent years have seen the emergence of trifluorothymidine and vidarabine, and no doubt other useful compounds of this class still remain to be discovered.

Some of the indications for the use of these compounds have long been established, such as the treatment of herpetic keratitis with idoxuridine. In other conditions their use is still experimental. For instance, there is as yet no satisfactory treatment of herpetic encephalitis, although it is reasonable to expect that success will eventually be achieved by sufficiently early treatment with one compound or another if a means of maintaining adequate concentrations in the cerebrospinal fluid can be devised. For this reason much attention has been paid to clinical pharmacology in the sections which follow. It will be noted that doses are sometimes expressed in mg/m². This is a common practice in cancer chemotherapy, and is based on the observation of Pinkel (1958) that generally accepted doses for patients in different age-groups which differ widely when expressed as mg/kg are essentially constant when expressed as milligrams per square metre of body area. This can be calculated from a formula given by Du Bois and Du Bois (1916); $\log S = 0.425 \log W + 0.725 \log H + 1.8564$, where S is surface area in square centimetres, W the weight in kilograms and H the height in centimetres. Nomograms for calculating the surface areas of children and adults based on this formula are given in Figure 3.1.

The naturally occurring nucleosides consist of a pyrimidine or purine base attached to a pentose sugar which is usually ribose or deoxyribose, or less commonly other pentoses. In those nucleosides used as antiviral agents the base is modified by chemical substitution, while the pentose sugar is ribose, deoxyribose or arabinose. Further information on the chemistry will be given in the sections which follow.

Before describing those nucleosides which are used in antiviral chemotherapy a brief account of the pathways of RNA and DNA synthesis will be given which will enable the mechanisms of action of the compounds to be understood.

The purine components are synthesized as their 5'-monophosphates. The starting material is 5'-phosphoribosyl-1-pyrophosphate. Various residues are transferred to this in successive stages from glutamine, glycine and N-formyltetrafolic acid to build up the purine ring system, the final product being inosine 5'-monophosphate, the riboside of hypoxanthine. This is then converted to adenosine 5'-monophosphate and guanosine 5'-monophosphate by two-stage processes, the amino groups coming from aspartic acid and glutamine respectively.

The pyrimidine components are built up in an entirely different manner. The starting materials are aspartic acid and carbamoyl phosphate. These combine and undergo ring closure, with one intermediate step, to give orotic acid. The phosphate moiety enters at this stage and is donated by 5'-phosphoribosyl-1-pyrophosphate. A decarboxylation step then occurs, giving uridine 5'-monophosphate. The cytosine analogue is formed from this at a later stage.

The purine and pyrimidine mononucleotides are converted to the di- and triphosphates by individual kinases. It is after this step that cytidine triphosphate is obtained from uridine diphosphate by the action of the enzyme cytosine triphosphate synthetase.

The four riboside triphosphates are incorporated into RNA by acting as substrates for RNA polymerase.

The nucleotides described so far all have ribose as the pentose moiety.

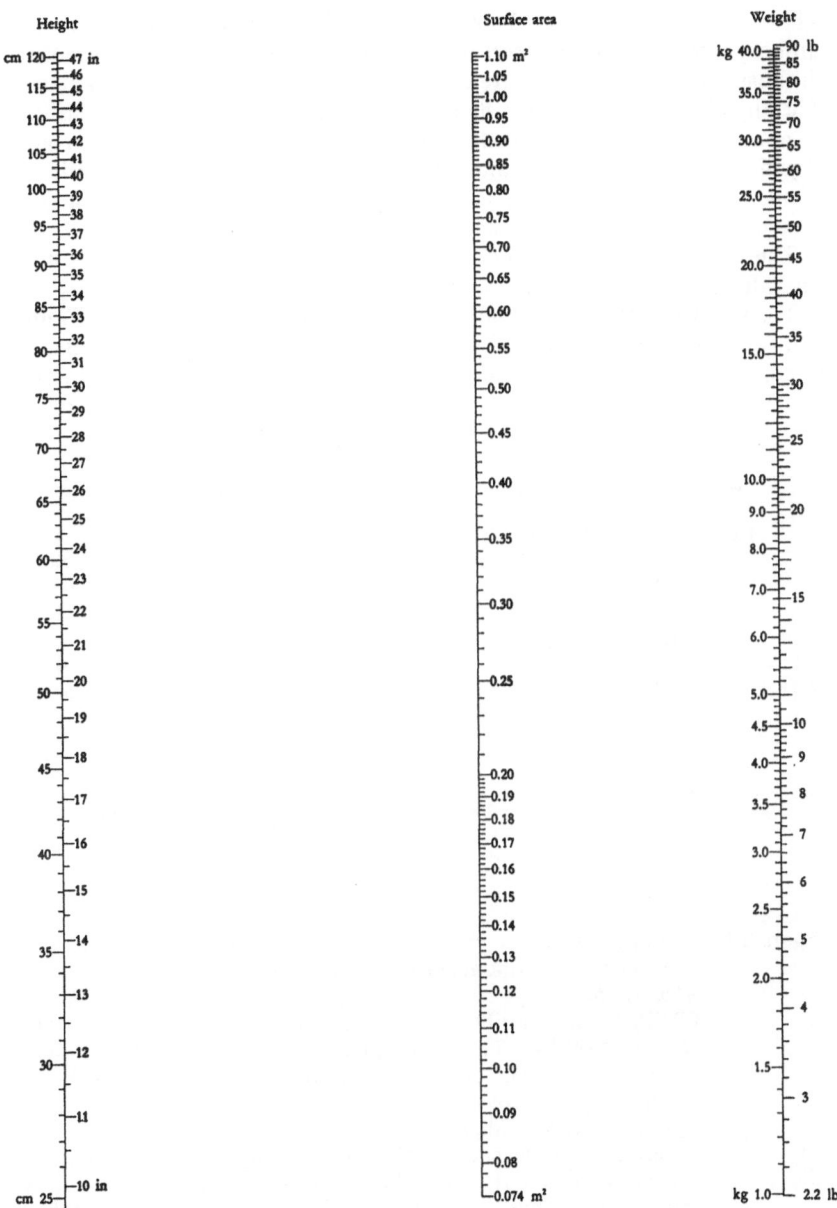

Figure 3.1 Nomograms for calculating surface area of body from height and weight (Documenta Geigy Scientific Tables, 1955)

The corresponding deoxyribosides are produced at the ribonuncleotide diphosphate stage as the result of reduction by the enzyme ribonucleotide diphosphate reductase, and are converted to the triphosphates by kinases. The thymine analogue arises by a different route, by which deoxyuridine 5'-monophosphate is methylated by thymidylic acid synthetase to give thymidine triphosphate. An alternative route exists by which thymidine is converted to the monophosphate by thymidine kinase. This is then converted to the di- and triphosphates by other kinases.

The four deoxyriboside nucleotides are incorporated into DNA under the influence of DNA polymerase.

The pathways of synthesis of RNA and DNA are summarized in Figure 3.2, which is an adaptation of one given by Davidson (1972). The sites of action of the antiviral nucleosides are indicated. Apart from the incorpora-

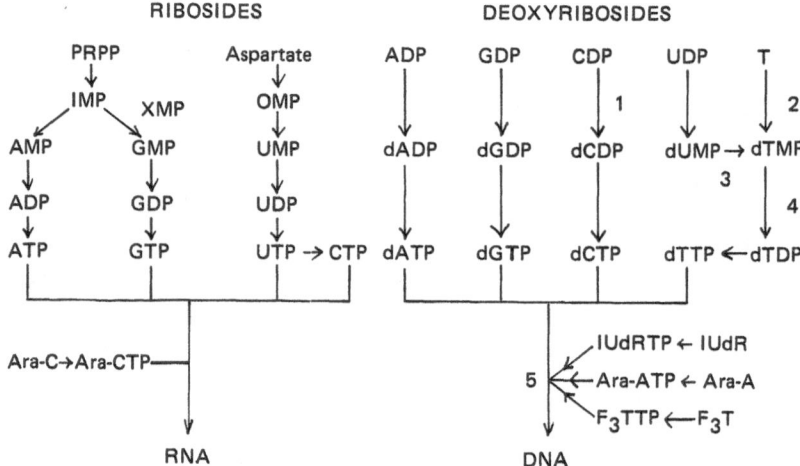

Figure 3.2   Pathways of nucleic acid synthesis and sites of action' of the antiviral nucleosides (modified from Davidson, 1972)
Abbreviations
PRPP, 5'-phosphoribosyl-1-pyrophosphate; MP, monophosphate; DP diphosphate; TP, triphosphate; A, adenosine; C, cytidine; G, guanosine; O, orotidine; T, thymidine; U, uridine; X, xanthosine; d, deoxy-; Ara-A, Vidarabine; Ara-C, Cytarabine; F_3T, trifluorothymidine; IUdR, Idoxuridine
Numerals indicate site of inhibition by antiviral nucleoside and enzyme inhibited. 1, Cytarabine, deoxycytidine kinase; 2, Idoxuridine and trifluorothymidine, thymidine kinase; 3, Cytarabine and trifluorothymidine, thymidylate synthetase; 4, Idoxuridine, thymidine monophosphate kinase; 5, Cytarabine, Idoxuridine and Vidarabine, DNA polymerase

tion of cytarabine into RNA, it will be seen that they all act on stages in the DNA pathway. As well as the enzymes coded for by herpes and vaccinia viruses, they also inhibit the same enzymes of the host cell; their antiviral action is thus non specific, and their clinical use depends upon the existence of a sufficiently high therapeutic ratio.

CYTARABINE

Cytarabine (Figure 3.3) is the generic name of 1-β-D-arabinofuranosyl-cytosine. It is often referred to in the literature as ara-C and cytosine arabinoside. It belongs to the chemical class of nucleosides, but contains the unnatural sugar arabinose instead of ribose or deoxyribose.

Cytarabine is a white crystalline powder, soluble in water to the extent of about 10% at room temperature. In aqueous solution it is hydrolytically deaminated by a first-order reaction to 1-β-D-arabinofuranosyluracil, which is biologically inactive (Notari, 1967), and this reaction is catalysed by phosphate and other buffer ions. Solutions of cytarabine should therefore be used within 48 hours of preparation.

Biological actions

Cytarabine was first brought into use as an antitumour compound. It was found to inhibit the growth of sarcoma 180 and Ehrlich carcinoma in mice treated intraperitoneally with daily doses of 5–50 mg/kg (Evans et al., 1961, 1964). It is used clinically in the treatment of acute myeloblastic leukaemia and acute and chronic lymphatic leukaemia, apart from its use as an antiviral agent.

As one of the consequences of its mode of action, which is to inhibit the synthesis of DNA, cytarabine has an immunosuppressive effect. Thus, the production of sheep haemolysin in rats could be completely suppressed by treatment with cytarabine in daily doses of 2000 mg/kg, a dose to which rats are resistant (Mitchell et al., 1969 a). Doses of 20 and 40 mg/kg daily will prolong allograft retention in mice (Griswold et al., 1972). Immunosup-pression has also been demonstrated in man (Mitchell et al., 1969 b). Patients with solid tumours who were under treatment with cytarabine showed a delayed primary immune response to injections of Escherichia coli Vi antigen and tetanus toxoid in comparison with control patients not receiving cytarabine, and in some cases the primary response was com-pletely inhibited. The antibody which was produced was sensitive to 2-mercaptoethanol and was thus presumably IgM. The secondary response was also delayed by cytarabine, but all patients produced antibody. Cytarabine also suppressed the induction of delayed hypersensitivity to 2,4-dinitrochlorobenzene. It has also been reported to cause chromosome breaks in cultures of human embryo lung cells (O'Neill and Rapp, 1971).

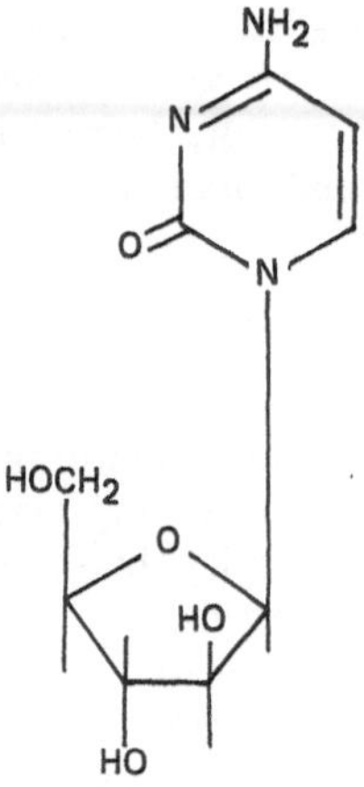

VIRUS DISEASES

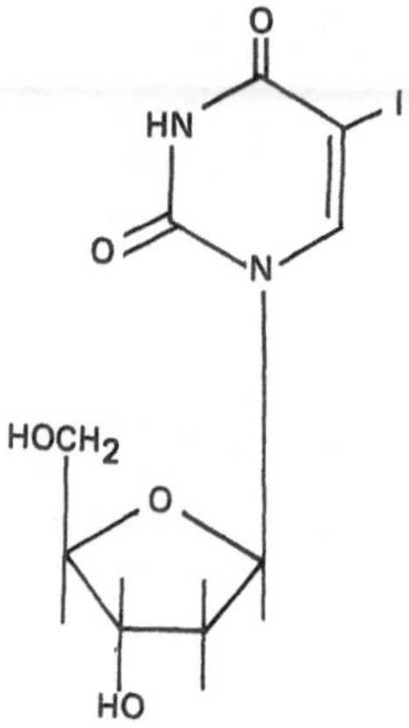

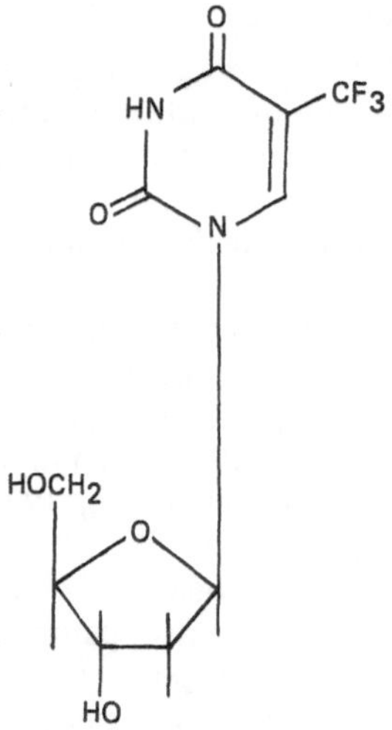

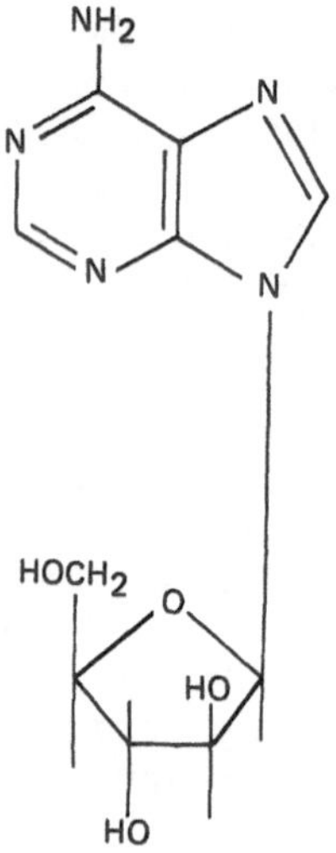

Figure 3.3 Structural formulae of the antiviral nucleosides: I, Cytarabine; II, Idoxuridine; III, Trifluorothymidine; IV, Vidarabine

Mode of action

Cytarabine is a cytotoxic agent. In studies carried out in cultures of mouse L cells cytarabine inhibited mitosis and DNA synthesis, but synthesis of RNA and protein was unaffected (Silagi, 1965). The effect was irreversible; most cells died, and the survivors became greatly enlarged. The cytotoxic effect could be partially prevented by deoxycytidine in a concentration 100 times that of cytarabine. Studies with labelled compound showed that cytarabine was incorporated into DNA, and to a lesser extent into RNA.

There has been a gradual change in views as to the precise biochemical basis of the action of cytarabine. In early work it was found that there was a 2–3-fold reduction in the content of deoxycytidine monophosphate in leukaemic cells, which was interpreted as indicating that cytarabine inhibited the synthesis of this compound (Chu and Fischer, 1962). This conclusion was supported by studies in ascites tumour cells (Kimball *et al.*, 1966), in which it was specifically shown that cytarabine blocked the conversion of cytidine monophosphate to deoxycytidine monophosphate by deoxycytidine kinase. Further work showed that cytarabine had additional actions. In ascites tumour cells it inhibited the incorporation of thymidine into DNA and inhibited DNA polymerase; this latter inhibition was reversed by deoxycytidine triphosphate, which is a substrate for the enzyme (Kimball and Wilson, 1968). Cytarabine is a competitive inhibitor of deoxycytidine kinase, which phosphorylates deoxycytidine to deoxycytidine monophosphate, and at the same time is converted by the enzyme to cytarabine monophosphate (Momparler and Fischer, 1968), and is converted further to the triphosphate (Schrecker and Urshel, 1968). DNA polymerase prepared from calf thymus and bovine lymphosarcoma was inhibited by cytarabine triphosphate, and this was considered to be the form in which the compound acts (Furth and Cohen, 1968). No incorporation of the triphosphate into DNA or RNA could be detected. In murine leukaemia cells, however, cytarabine was found to be incorporated into both DNA and RNA, and incorporation into terminal positions of RNA strands was postulated as the cause of cell death (Chu and Fischer, 1968). Experiments *in vitro* with DNA polymerase partially purified from calf thymus showed that cytarabine was incorporated into DNA, mainly at the 3′-hydroxyl terminus. This might block further extension of DNA chains, leading to cell death for this reason (Momparler, 1969). This view received apparent support from experiments in which L cells were exposed to labelled cytarabine for a short period just sufficient to allow DNA to form, since the labelled compound was found in short pieces of DNA. However, when cytarabine was removed and deoxycytidine was added to permit DNA synthesis to continue, the DNA strands continued to grow, and degradation studies with nucleases indicated that most of the cytarabine was not in terminal positions (Graham and Whitmore, 1970). Cytarabine therefore did not block elongation of the DNA strands, and its inhibitory effect on DNA synthesis could be accounted for by inhibition of DNA polymerase. This is probably the main mechanism of action, since studies on the ribonucleotide diphosphate reductase stage have indicated that inhibition of this enzyme is not the primary site of action of cytarabine (Inagaki *et al.*, 1969); inhibition of thymidylate synthetase requires higher concentrations of cytarabine than those capable of inhibiting thymidine incorporation into DNA, and is therefore only a secondary action of cytarabine (Roberts and Loehr, 1972). Also, the nucleotides of cytarabine were found to be only weak inhibitors of the reduction of ribo- to deoxyribonucleotides by ribonucleotide diphosphate reductase, so that this action is probably not important in inhibition of DNA synthesis (Moore and Cohen, 1967).

The pathways involved in the inhibition of DNA synthesis by cytarabine are shown in Figure 3.2. Cytarabine competitively inhibits deoxycytidine kinase, thus reducing the supply of deoxycytidine monophosphate to nucleoside diphosphokinase. There is thus less deoxycytidine triphosphate available for incorporation into DNA, and this final stage is further reduced by inhibition of DNA polymerase. The involvement of cytarabine in RNA synthesis has not been studied in such detail. It does not inhibit RNA polymerase, but there is evidence to suggest that it is incorporated, and one may well suppose that RNA containing cytarabine residues may be defective in function. Chu and Fischer (1968) found that the inhibition of DNA polymerase activity caused by cytarabine was reversible on removal, but the cells remained non-viable. Cell death caused by cytarabine must therefore be due to some other cause than inhibition of DNA synthesis. It was further shown (Chu, 1971) that cytarabine incorporated into RNA appeared in species of low molecular weight (2–16S), and when this happened incorporation of uridine into RNA was prevented. Cell death may therefore be due to inhibition of the synthesis of low molecular weight RNA, which is so far the most basic biochemical site of action to have been established.

Antiviral action

Since the main action of cytarabine is to inhibit DNA synthesis it was a logical consequence to investigate its effect upon the multiplication of DNA viruses. It was found to inhibit the formation of plaques by vaccinia virus in embryo chick and rat kidney cells, and this action was presumably intracellular since the compound did not inactivate the virus on contact and did not prevent its uptake by the cells (Renis and Johnson, 1962). However, the antiviral effect of cytarabine was observed at concentrations not much below toxic levels (Buthala, 1964), and was therefore possibly due to non-specific inhibition of both cell and virus DNA synthesis. Nevertheless, cytarabine was found to be effective in the treatment of experimental herpetic keratitis in rabbits. When administered in repeated doses as 1% eye drops or 1% ointment the herpetic ulceration was reduced or eradicated in comparison with the clinical course in untreated control animals (Underwood, 1962). A marked reduction in the extent of the lesions was also observed in rabbits after 2-hourly treatment with 0·1% eye drops beginning on the 4th day of infection. A similar result was also obtained in squirrel monkeys, in which the anatomy of the cornea is more typical than in the rabbit (Kaufman and Maloney, 1963 a).

Cytarabine was also found active in tissue culture tests against other members of the herpes virus group, including varicella-zoster (Rapp, 1964), pseudorabies and B virus (Buthala, 1964), herpes saimiri, a virus causing malignant lymphoma and lymphocytic leukaemia in owl monkeys (Adamson et al., 1972), and cytomegalovirus (Sidwell et al., 1972; Váczi and Gönczöl, 1973). Most of the work on the chemotherapy of herpes virus infections has been carried out with poorly characterized strains of virus. It has been known for some time that strains of human herpes virus can be differentiated into two types, type 1 causing mainly eye and skin infections, and type 2 causing the majority of genital infections. In view of the prevalence of type 2 infections in venereal practice and their possible association with carcinoma of the cervix it is of importance to determine the chemotherapeutic sensitivity of these strains in comparison with type 1. There seems to be general agreement that type 2 strains are less sensitive to cytarabine. Thus, in tissue culture tests carried out under defined conditions the minimum inhibitory concentration for seven type 1 strains was 0·25–0·75 µg/ml, and for four of nine type 2 strains it was greater than 0·9 µg/ml (Fiala et al., 1972). In similar work with a wider range of viruses the minimum inhibitory concentrations were again 0·25–0·75 µg/ml for herpes type 1, 0·37–2·0 µg/ml for herpes type 2, 2–4 µg/ml for cytomegalovirus and 0·37–4·0 µg/ml for varicella-zoster. Cytarabine was more active in these tests than vidarabine and idoxuridine (Fiala et al., 1974).

Other members of the poxvirus group besides vaccinia are also sensitive to cytarabine, such as swinepox and fowlpox (Buthala, 1964) and Shope fibroma (Minocha and Maloney, 1970). There appear to be no reports on its activity against variola, but cytarabine has been used with apparent success in the treatment of smallpox (Hossain et al., 1972).

Some types of adenovirus are inhibited by cytarabine, but not all. Cytarabine has been reported active in tissue culture against types 2, 7, 12, 19 and the simian adenovirus SV15 (Feldman and Rapp, 1966; Fong et al., 1968), but not against types 1, 5 and 7, and in addition the activity claimed against type 2 was not confirmed (Herrmann, 1968). There have been no reports of the clinical use of cytarabine in adenovirus infections.

Cytarabine will inhibit the multiplication of SV40 virus, an oncogenic simian papovavirus occurring as a contaminant in early batches of poliomyelitis vaccine (Rapp *et al.*, 1965), fibroma virus (Roby *et al.*, 1965), and also H-1 virus, a parvovirus of rodents (Ledinko, 1967).

The replication of a number of RNA viruses is also inhibited by cytarabine. A concentration of $10^{-3\cdot5}$ M gave almost complete inhibition of the multiplication of Rous sarcoma and Rous-associated viruses in chick embryo cells (Bader, 1965). Multiplication was also inhibited by actinomycin D and mitomycin C. These results indicate that a stage of DNA synthesis is required in the multiplication of these two RNA viruses. A similar situation was observed with Friend leukaemia virus (Yoshikura, 1968); multiplication in mouse lung cells was almost completely inhibited by 100 µM cytarabine, and the same concentration produced 95% inhibition of the incorporation of labelled thymidine into DNA; here again it appears that a transitory stage of DNA synthesis forms part of the virus growth cycle. Cytarabine in a concentration of 10 µM also inhibits the multiplication of Moloney sarcoma virus (Hirschman, 1969), and a requirement for a DNA replication step is thereby implied.

Cytarabine inhibits the formation of virus antigens in BHK-21 cells infected with rabies virus (Campbell *et al.*, 1968). The inhibition can be reversed by cycloheximide; an inhibitor of protein synthesis at the translational stage. It is therefore possible that the antiviral action of cytarabine depends upon the induction of a cell protein. A similar effect has been observed with isatin 3-thiosemicarbazone (an analogue of methisazone), since the inhibition of the multiplication of rabbitpox virus which it produces in HeLa cells is reversed by actinomycin D in concentrations which are too low to have any antiviral effect of their own (Appleyard *et al.*, 1965). It is also possible that incorporation into RNA may be concerned in the inhibition of RNA virus antigens by cytarabine.

The antiviral spectrum of cytarabine is shown in Table 3.1. Its clinical use has so far been limited to the treatment of herpes keratitis and encephalitis, smallpox and various manifestations of varicella-zoster infection.

Toxicity

The cytotoxicity and immunosuppressive activity of cytarabine have been briefly mentioned in the section on biological activity. The cytotoxic activity is quite high, since in a standard 3-day cytotoxicity test the ID50 for cytarabine in KB cells was 0·03–0·1 µg/ml. The toxicity could be reversed by deoxycytidine in a concentration of 10 µg/ml (Smith *et al.*, 1969). The viability of HeLa cells was greatly reduced after exposure to 8·2 µM cytarabine for one generation time of 22 hours. Inhibition of DNA synthesis with continued synthesis of RNA and protein was considered to give rise to a condition of unbalanced growth leading to cell death (Kim and Eidinoff, 1965). As a consequence of its cytotoxic action cytarabine will cause chromosome breakage. After exposure to a concentration of

Table 3.1 Reported spectrum of antiviral activity of cytarabine

Type	Group	Virus
DNA	Herpesvirus	Herpes 1
		Herpes 2
		Pseudorabies
		B virus
		Herpesvirus saimiri
		Varicella-zoster
		Cytomegalovirus
	Poxvirus	Vaccinia
		Swinepox
		Fowlpox
		Fibroma
	Adenovirus	Adenovirus 2 (?)
		Adenovirus 7
		Adenovirus 12
		Adenovirus 19
		SV 40
	Parvovirus	H-1
RNA	Rhabdovirus	Rabies
		Vesicular stomatitis
	Leukovirus	Moloney murine sarcoma
		Harvey murine sarcoma
		Friend leukaemia
		Rous sarcoma

3×10^{-5} M for 4 hours breaks were observed in 86% of metaphases in experiments with human leukocytes (Kihlman et al., 1963). Chromosomal bridges, fragmentation and polyploidy have been seen in preparations from patients treated with cytarabine (Talley et al., 1962).

Cytotoxic effects can also be demonstrated in vivo, both in animals and in man. In rats given cytarabine intraperitoneally in a single dose of 250 mg/kg pathological changes could be observed in the intestine 32 hours later, consisting of necrosis of the crypt cells and cells of the germinal centres of the Peyer's patches (Liebermann et al., 1970). This toxic effect could be entirely prevented by 1·5 mg/kg cycloheximide given intraperitoneally. It was concluded that cell death ensues when DNA synthesis is inhibited, but not when protein synthesis is inhibited simultaneously. The ultimate cause of the cell death produced by cytarabine might therefore be a toxic protein, which is not produced in the absence of the compound. Cytarabine inhibits marrow regeneration in lethally irradiated mice recolonized by donor marrow cells. This effect is obtained with a single intraperitoneal dose of 100 mg/kg given on the 4th day after irradiation (Papac *et al.*, 1965). A teratogenic effect has been observed in fertile eggs injected with cytarabine into the yolk sac on the 4th day of incubation. The LD50 was around 0·025 mg per egg. Embryos which survived to 18 days were stunted, with facial coloboma, absence of pelvic skeleton and other bone deletions, corneal cysts and inhibition of feather growth. Embryos surviving from eggs treated after the 8th day of incubation showed only reduction in body weight, disturbances of feather growth and cerebellar atrophy (Karnofsky and Lacon, 1966). Cytarabine has also been reported to be mutagenic (Huberman and Heidelberger, 1972).

The extensive clinical use of cytarabine in the treatment of malignant conditions has provided abundant evidence of its toxic effects in man. In a typical study 13 patients were given cytarabine intravenously in a dose of 3–10 mg/kg daily for 4–9 days to a total dose of 50 mg/kg, or single doses of 30–50 mg/kg at 7–10-day intervals. Reductions in haemoglobin, red cells and platelets were observed. The bone marrow showed megaloblastic changes, with giant metamyelocytes and megaloblasts. The cells showed sticking chromosomes in anaphase, tripolar distribution of chromosomes, poor alignment of mitotic figure, and increased number of chromosomes leading to polyploidy. The peripheral blood contained hypersegmented neutrophils, macrocytes, and megaloblasts with three nuclei and also micronuclei (Talley and Vaitkevicius, 1963). Nausea and vomiting are major side effects, and are more frequent with an intermittent schedule of dosing than with continuous infusion; the onset is usually 2–4 hours after injection and the duration 1–4 hours. Less common side effects include conjunctivitis and keratitis, mucosal ulceration and lethargy and confusion (Talley *et al.*, 1967).

The corneal toxicity of cytarabine is important in view of its use in treating herpetic keratitis. Kaufman *et al.* (1964) treated the eyes of healthy volunteers with eye drops of 0·1%, 0·5% and 1·0% cytarabine, given six times a day for 7 days. On slit lamp examination glittering opacities could be seen in the lower layers of the corneal epithelium, and there was punctate staining with fluorescein. There was some pain, and iritis also developed. The corneal deposits appeared after an interval of 4–7 days, and disappeared over a period of 3 weeks after treatment was discontinued. In an investigation in rabbits similar lesions were seen; lactic and α-glycerophosphate dehydrogenase disappeared from the superficial layers of the cornea and megalocytes were seen in the basal layers. Similar effects were described by Elliott and Schat (1965).

Metabolism

The conversion of cytarabine to the triphosphate has already been described, and also its incorporation into DNA and RNA. It is also broken down into inactive products by metabolic activity. When incubated with homogenates of human liver and kidney it is metabolized to the uracil analogue by pyrimidine nucleoside deaminase (Camiener and Smith, 1965), a compound which is devoid of antiviral activity.

Clinical pharmacology

The clinical pharmacology of cytarabine has been extensively studied in the course of its use in the treatment of malignant conditions. When the tritium-labelled compound was given intravenously radioactivity in the plasma declined with a half-life of 30–60 minutes. Excretion of radioactivity in the urine was 50% complete after 4–8 hours and almost complete by 24 hours; 89–96% of the radioactivity was accounted for by the inactive uracil analogue produced by enzymatic deamination (Creasey et al., 1964). Similar results were obtained on administration by other routes. Excretion in the urine was rapid, and at 24 hours the recovery of radioactivity after subcutaneous administration was 70%, intramuscular 80% and intrathecal 71%. The in vivo half-life, as distinct from the plasma half-life, could be calculated from the urinary excretion, and was $6\frac{1}{2}$–8 hours, $2\frac{1}{2}$–6 hours and $8\frac{1}{2}$ hours by the three respective routes, and 3–5 hours after intravenous administration (Finkelstein et al., 1970). Other workers have found a shorter plasma half-life, and the actual value may depend upon the general condition of the patients used in the study. Baguley and Falkenhaug (1971) found a half-life of 3–9 minutes, and Ho (1971) found that the plasma disappearance curve was biphasic, with an initial phase of rapid disappearance with a mean half-life of 12 minutes, and a slower second phase of mean half-life 111 minutes. When given intrathecally the half-life is longer (2 hours), on account of the lower deaminase activity of the brain and cerebrospinal fluid.

The dosage schedules used in antiviral chemotherapy have been based on those worked out for the treatment of malignant conditions. Both rapid and continuous intravenous infusion are used, and daily doses range from 0·3 to 8 mg/kg and from 10 to 100 mg/m². The schedules will be given in detail in the chapters on the treatment of the individual virus diseases.

Assay methods

Several methods have been used for determining the concentration of cytarabine in body fluids, and all depend upon utilization of the biological properties of the compound.

There are no reports of spectrophotometric methods, probably because the main absorption of cytarabine is in the ultraviolet region, where other substances may cause interference.

(i) A method based on the antiviral activity of cytarabine has been described by Buthala (1964). Samples of serum or plasma are mixed with an equal volume of acetone and incubated at 45 °C for 20 minutes. The proteins are precipitated, and herpes antibody, which would interfere with the test, is thus removed. The precipitate is removed by centrifugation and the acetone is boiled off. An appropriate dilution of the aqueous solution thus obtained is mixed with overlay medium and placed on a cell monolayer infected with sufficient herpes virus to produce a countable number of plaques. After incubation for 4 days at 37 °C the size of the plaques is determined by measurement under a low power microscope, and the mean diameters read off against a standard curve obtained with known concentrations of cytarabine. The method is sufficiently sensitive to determine concentrations of $0\cdot1-0\cdot5$ µg/ml in the overlay medium, and it can also be applied to tissue extracts.

(ii) Cytarabine can be determined by its cytotoxic effect on KB cells. This method was first applied to the determination of cycloheximide and streptovitacin (Smith et al., 1959), and subsequently used in a study of the clinical pharmacology of cytarabine by Talley et al. (1967). Aliquots of 4 ml of a suspension of KB cells in culture medium are pipetted into tubes containing $0\cdot1-0\cdot2$ ml of the test samples, and the tubes are incubated at 37 °C for 72 hours. The medium is then removed and the cell sheets are washed several times and then dissolved in biuret reagent. The protein content is then determined by spectrophotometry in comparison with standards prepared with crystalline bovine albumin. Control cultures not containing test samples are treated similarly, and the percentage inhibition of protein synthesis produced by the sample can then be calculated. The concentration of cytarabine can then be determined by comparison with standards set up with known concentrations of the compound. The sensitivity of the method is around $0\cdot5$ µg/ml.

(iii) Cytarabine will inhibit the growth of the ATCC 8043 strain of *Streptococcus faecalis*, and this effect has been used as the base for an assay (Hanka et al., 1970). Plates containing Difco folic acid medium with $1\cdot5\%$ agar, $0\cdot5$ µg/ml folic acid and 20 µg/ml tetrahydrouridine are seeded with the organism and discs dipped in the test samples are placed on the surface. Incubation is continued for 18–25 hours at 37 °C. The concentration of cytarabine is determined from the zone of inhibition with the aid of a calibration curve. The sensitivity of the method is $0\cdot1$ µg/ml.

(iv) Cytarabine reduces the rate of multiplication of L cells, and an assay utilizing this effect has been developed by Borsa et al. (1969). Suspension cultures are rolled at 37 °C for 20 hours, by which time the cells are in the exponential growth phase. The test samples are then added to the tubes and incubation is continued for 48 hours. The cells are then counted, and the count is expressed as a multiple of the count at 20 hours. The maximum increase in count in the absence of cytarabine is around 5-fold, and the cytarabine content of the samples can be calculated by reading off the cell count against a calibration curve obtained with known concentrations. The sensitivity of the method is not stated explicitly, but a 50% reduction in cell growth was produced by a concentration of 0·01 µg/ml.

(v) Baguley and Falkenhaug (1971), in their study of the clinical pharmacology of cytarabine, developed an improved method based on inhibition of the uptake of labelled thymidine into DNA. The sensitivity is 0·04 µg/ml and the determinations can be carried out in 4 hours, a valuable feature when the treatment of a patient is being monitored. Plasma is obtained by taking 5 ml samples of blood into heparinized tubes containing 50 µg of tetrahydrouridine, which inhibits enzymatic degradation of cytarabine. A 1 ml sample of plasma is mixed with 0·05 µCi of tritiated cytarabine dissolved in 50% ethanol, and applied to a column of Sephadex G-10. The column is eluted with successive 0·7, 0·8 and 0·8 ml portions of medium, and the fractions collected are tested for radioactivity. The middle fraction contains almost no protein and 50–70% of the radioactivity, and thus presumably the same percentage of the amount of cytarabine originally present in the sample. This step could now be more conveniently replaced by small-volume equilibrium dialysis. The protein-free samples are then added to suspensions of L cells or mouse spleen cells in a medium containing 1 µCi/ml of tritiated thymidine. After incubation at 37 °C for 2 hours the mixtures are cooled to 4 °C, mixed with an equal volume of 0·3% Nonidet P-40 in 0·9% saline and shaken. The cell suspensions thus obtained are washed on to glass fibre filter discs and washed further with water, 1% and 5% trichloroacetic acid. The filter discs are then counted in a scintillation counter. The radioactivity recorded is proportional to the incorporation of thymidine into DNA, and can be expressed as a percentage of the incorporation in control cell suspensions not treated with cytarabine. The actual concentration of cytarabine is then read off from a reference line obtained with known concentrations of the agent. Apart from rapidity and sensitivity, the method also has the advantage that it does not necessarily require tissue culture facilities.

The various methods for determining cytarabine are summarized in Table 3.2. They all require the availability of bacteriological or tissue culture facilities, and cannot be regarded as coming within the category of routine investigations.

Clinical use

Cytarabine is used in the treatment of infections caused by the herpesvirus group, and there is one report of the successful treatment of smallpox

Table 3.2 Methods for determining cytarabine

Author	Method	System	Time required	Sensitivity (μg/ml)
Buthala, 1964	Antiviral action	Tissue culture	4 d	0·1–0·5
Smith et al., 1959	Cytotoxicity	KB cells	3 d	0·5
Hanka et al., 1973	Antibacterial action	Str. faecalis	1 d	0·1
Borsa et al., 1969	Growth inhibition	L cells	3 d	0·01
Baguley and Falkenhaug, 1971	Inhibition of DNA synthesis	L cells	4 h	0·04

(Hossain et al., 1972). The indications for using cytarabine resemble those for idoxuridine and vidarabine, and are shown in Table 3.3, together with the route of administration and type of preparation used. Unlike idoxuridine, cytarabine has not been used as a solution in dimethylsulphoxide for topical use.

Contraindications

In view of its immunosuppressive and cytotoxic actions cytarabine should be administered with caution to patients with depression of bone marrow function resulting from irradiation and the use of cytotoxic drugs. Its use in pregnancy is inadvisable.

Preparations

Cytosar (Upjohn). Ampoules containing 100 mg of cytarabine and 5 ml of solvent consisting of water for injection containing 0·9% benzyl alcohol. A solution of the compound may be infused into an intravenous drip, and the solid may be used for the preparation of eye-drops.

IDOXURIDINE

Idoxuridine (Figure 3.3) is the generic name of 5-iodo-2'-deoxyuridine. It is often referred to as IUdR (iodouracil deoxyriboside). It is a synthetic nucleoside containing the natural sugar deoxyribose.

Idoxuridine is weakly acidic (pK 8·25), soluble in water but unstable when heated in solution. In 5% glucose or 0·45% sodium chloride the maximum solubility at room temperature is 8 mg/ml, and solutions are usually prepared at 5 mg/ml to avoid crystallization on storage. Solutions

Table 3.3 Clinical use of cytarabine

Group	Virus	Condition	Route	Preparation
Herpesvirus	Herpes	Keratitis	Topical	Eye drops
		Cutaneous herpes Generalized	Intravenous	Aqueous solution
		Eczema herpeticum	Intravenous	Aqueous solution
		Buccal gingivostomatitis	Topical	5% in Orabase
		Encephalitis	Intravenous ⎱ Intrathecal	Aqueous solution
			Intravenous	Aqueous solution
	Varicella-zoster	Pneumonia	Intravenous	Aqueous solution
		Zoster Generalized Varicella In immunosuppressed patients	Intravenous	Aqueous solution
	Cytomegalovirus	Cytomegalic inclusion disease	Intravenous	Aqueous solution
Poxvirus	Vaccinia Variola	Keratitis	Topical	Eye drops
		Smallpox	Intravenous	Aqueous solution
Papovavirus		Progressive multifocal leukoencephalopathy	Intravenous ⎱ Intrathecal	Aqueous solution

cannot be autoclaved and must be sterilized by filtration. These unfavourable properties restrict the clinical usefulness of the compound (Calabresi, 1963).

In unbuffered aqueous solution (pH 5–6) idoxuridine is decomposed by ultraviolet light and less rapidly by visible light. Solutions should therefore be kept in dark glass bottles. They have a reasonable storage life, and decomposition to the extent of only 2% has been observed after 3 months at room temperature (Simpson and Zappala, 1964).

Determination of the rates of hydrolysis in aqueous solution between pH 1·3 and pH 12·0 shows that idoxuridine is fairly stable between pH 2 and pH 7, and there is a plateau of considerably lesser stability between pH 9 and pH 11. At acid pH idoxuridine is hydrolysed to iodouracil and deoxyribose, and thence to uracil and iodide ion. At alkaline pH the degradation proceeds mainly via 5-hydroxy-2′-deoxyuridine to deoxyuridine and uracil. In both routes antiviral activity is lost at the first step in hydrolysis (Garrett et al., 1964, 1965; Ravin et al., 1964).

Idoxuridine is soluble in dimethylsulphoxide to the extent of at least 40%, and such solution are made up at the time of use for topical application. The solvent should be of spectrographic grade.

Solutions for intravenous infusion are made up at the time of use, and the procedure of Tomlinson, as reported by Conchie et al. (1968), may be employed. The recommended concentration is 0·5% w/v. The required volume of dextrose injection (B.P., B.N.F.; 5% w/v) is adjusted to pH 10 with solid sodium carbonate and heated to 37–40 °C in a vessel provided with a mechanical stirrer. The idoxuridine is finely ground and added slowly; the pH falls as it passes into solution, and is adjusted to pH 10 with sodium bicarbonate. When all is dissolved the pH is adjusted to 9 with sodium carbonate or acid as necessary, and the solution is sterilized by filtration. It may be stored at 4 °C for not more than 14 days.

Biological actions

In early studies of idoxuridine it was found to be cytotoxic. In the presence of concentrations ranging from 10 μM to 100 μM the cells of the murine leukaemia lymphoblast line L 5178 completed only one doubling before cell death occurred (Mathias et al., 1959). On the basis of these and similar results its effect was tried in malignant conditions in màn (Calabresi et al., 1960). Some degree of arrest or regression of the tumours was noted with daily doses of 100–120 mg/kg for 4–7 days, but idoxuridine has not turned out to be of established value in this field.

In mice immunized with sheep red cells and treated with 2 mg idoxuridine intraperitoneally on 4 successive days the number of antibody-forming cells in the spleen as determined by the Jerne method was markedly reduced in comparison with untreated control animals. Idoxuridine thus has an immunosuppressive effect (v. Thiel et al., 1967).

Idoxuridine has not been reported to be carcinogenic, and it did not act as an initiator when given in conjunction with croton oil for the induction of skin tumours in mice (Trainin *et al.*, 1964).

Idoxuridine is a potent teratogenic agent in mice (Skalko and Packard, 1973). Pregnant females were given single intraperitoneal injections of 100–500 mg/kg on days 7–11 of pregnancy, which cover the period of organogenesis, and killed on day 17. Doses of 300 mg/kg and 500 mg/kg produced 45–100% fetal resorptions, according to the day of pregnancy on which they were given. The survivors showed a marked incidence of exencephaly, polydactyly, cleft palate, omphalocele and ectrodactyly. Doses of 100 mg/kg or lower did not produce any ill effects.

Although idoxuridine is an antiviral agent, it can facilitate virus multiplication in certain circumstances. In work reported by St. Jeor and Rapp (1973 a, b) human embryo lung fibroblasts were incubated with 100 µg/ml idoxuridine for 72–96 hours and then washed and soaked in buffer for 4 hours to remove any intracellular idoxuridine which had not been incorporated into DNA. They were then infected with cytomegalovirus, and in comparison with cells infected without pretreatment the eclipse phase was shortened and the yield of virus per cell was increased 5-fold. It was suggested that idoxuridine prevented the synthesis of a substance present in normal cells which inhibits the multiplication of the virus. Cytomegalovirus normally grows only in cultures of human fibroblasts, although it is frequently isolated from epithelial tissue. When human embryo kidney cells were grown for 96 hours in the presence of idoxuridine and then washed, it was found that they were then able to support multiplication of the virus to a high titre. The compound thus converted the cells into a permissive state, again presumably by preventing the formation of a normally-occurring inhibitory substance.

A similar effect may underly the induction of tumour viruses which can be brought about by idoxuridine. Cells of embryo mice of the high leukaemic strain AKR can be grown in culture as virus-negative cell lines. Exposure of these cells to 20 µg/ml or 100 µg/ml idoxuridine for 24–48 hours causes murine leukaemia virus to appear in 0·5% of the cells within 7 days. The activation of the virus appears to require incorporation of idoxuridine into DNA, since the effect is inhibited by equimolar thymidine (Lowry *et al.*, 1971). A similar result has been obtained with Epstein–Barr virus, a member of the herpesvirus group associated with Burkitt's lymphoma. Cells of the Burkitt lymphoblastoid cell line P3J-HR-1 were fused with Sendai virus to a human sternal marrow cell line. Monolayers of these hybrid cells were incubated with 40 µg/ml idoxuridine for 3 days, washed, and incubated for a further 3 days in the absence of the compound. The cells now contained Epstein–Barr virus, which could be detected by indirect immunofluorescence microscopy, and by the examination of sections in the electron microscope (Glaser and Rapp, 1972).

There have been no reports of induction or facilitation of virus occurring during the clinical use of idoxuridine, but it would be advisable to bear in mind the possibility, however unlikely, if treatment with idoxuridine is to be prolonged, or if the infection is not responding to treatment.

Mode of action

Idoxuridine is an analogue of thymidine, differing from it in having an iodine atom in the 5-position instead of a methyl group. This substitution slightly increases the atomic radius, but not to an extent sufficient to hinder the incorporation of idoxuridine into DNA. It reduces the pK_a from 9·8 to 8·25, with the consequence that the proportion of the molecule in the enolic form, in which hydrogen bonding can occur, is 35 times greater than in thymidine, thus affording the possibility of errors in transcription and chain separation (Prusoff, 1963).

Incorporation of idoxuridine into DNA was first demonstrated by Eidinoff *et al.* (1959). HEp-1 cells were grown in the presence of $1·4 \times 10^{-4}$ M idoxuridine labelled with ^{131}I; the DNA was then isolated and was found to contain the radioactive label. Determination of the proportions of the bases indicated that idoxuridine had replaced thymidine.

Mathias and Fischer (1962) found that leukaemia lymphoblasts died when exposed to idoxuridine in a concentration of 10 μM. DNA isolated from the cells contained idoxuridine to the extent of $\frac{1}{3}$ to $\frac{1}{2}$ of the amounts of the normally occurring deoxyribonucleotides and this incorporation was presumably the cause of the cytotoxic effect.

In addition to being incorporated itself, idoxuridine reduces the incorporation of [^{3}H]thymidine into DNA. The site affected depended upon the tissue being examined. In Ehrlich ascites carcinoma cells and human chronic granulocytic and acute monocytic leukaemias idoxuridine inhibited DNA polymerase, but in murine leukaemia cells it inhibited thymidine kinase and thymidylic acid kinase (Delamore and Prusoff, 1962). Idoxuridine inhibits the thymidine kinase of BSC-1 cells, and also the thymidine kinase induced by herpesvirus growing in the same cells (Prusoff *et al.*, 1965).

In spite of inhibition of enzymes of the DNA pathway, the synthesis of DNA is not prevented. Morris and Cramer (1966) found that cultures of murine mast cells underwent only one doubling of cell number in the presence of 100–200 μM idoxuridine. Studies on the uptake of labelled thymidine showed that DNA was still being synthesized, but was breaking down at such a rate that there was no net accumulation. The condition of the cells did not revert to normal when they were transferred to a medium not containing idoxuridine. The compound was incorporated into DNA, and this mixed DNA was able to replicate to some extent.

Idoxuridine will inhibit the multiplication of herpesvirus, as described in the following section. In human embryo fibroblasts and primary rabbit kidney cell cultures a concentration of 100 µg/ml was sufficient to prevent the formation of infective virus, but a reduced yield of imperfect particles could be seen in the cells on electron microscopy. Formation of virus components still took place, since they could be detected by immunofluorescence microscopy, and idoxuridine may exert its antiviral action by interrupting the assembly of components into mature virions (Smith and Dukes, 1964). Similar results were obtained by Kaplan and Ben-Porat (1966) in rabbit kidney cells infected with pseudorabies virus, a virus of the herpes group. Idoxuridine was incorporated into the virus DNA to the extent of replacing 90% of the thymidine. The protein components of the virus were still synthesized, but they were not incorporated into virions. If idoxuridine was withdrawn from the cultures the virus DNA could still be incorporated to the extent of becoming inaccessible to DNAse. The lack of formation of infective virus was therefore not due to distortion of the DNA molecule by the idoxuridine which it contained, but probably resulted from the inability to produce some protein involved in assembly of the virion.

The known biochemical activities of idoxuridine are summarized in Figure 3.2, but it is clear that the specific biochemical defects underlying the mode of action of idoxuridine have not yet been elucidated, and further studies of the formation of virus protein in the presence of idoxuridine are required, in order to detect absence of particular components, deficiencies in amounts or qualitative changes (Prusoff and Goz, 1973).

Antiviral activity

The antiviral activity of idoxuridine was first reported by Herrmann (1961), who found that it was active against vaccinia and herpes viruses in plaque inhibition tests, but inactive against West Nile and Newcastle disease viruses. These findings are in accordance with the involvement of idoxuridine in DNA synthesis, and they soon led to the use of idoxuridine as a reagent for the determination of the nucleic acid type of viruses which had not yet been characterized. Thus Hermodson and Dinter (1962) found that the growth of pseudorabies virus in calf kidney cell cultures was inhibited by idoxuridine, and it was therefore inferred to be a DNA virus. Bovine virus diarrhoea virus and a bovine enterovirus were not inhibited, and were therefore RNA viruses. Susceptibility to inhibition by idoxuridine also enabled feline rhinotracheitis virus (Ditchfield and Grimyer, 1965) and African swine fever virus (Plowright et al., 1965) to be identified as DNA viruses.

Shortly after the discovery of the activity of idoxuridine against herpes virus in tissue culture Kaufman (1962) was able to show that it was active in vivo, and thus of clinical interest. Rabbits inoculated with herpes virus on the scarified cornea developed severe dendritic keratitis and iritis; the infection was well established after 24 hours and progressed for at least 10 days. No infection developed if the eyes were treated with a saturated solution of idoxuridine given 2-hourly, beginning before infection or up to 12 hours afterwards. When treatment was begun 48 hours after infection a complete cure was obtained after 48 hours of treatment in six of six eyes treated. Cure was also obtained in all of three eyes in which treatment was begun on the 5th day of infection. The infection did not recur after the end of treatment. This result was confirmed by Perkins *et al.* (1962) in similar experiments, and Kaufman and Maloney (1963 a) made the additional observation that treatment was still effective when involvement of the deeper layers of the cornea was produced by deep scarification. The activity of idoxuridine was further confirmed in a trial carried out under double-blind conditions (Corwin *et al.*, 1963). Herpetic keratitis was produced in 16 rabbits, and 48 hours after infection 0·1% eye drops of idoxuridine or a placebo solution were given 2-hourly for 5 days. A cure was obtained in ten of 16 eyes treated with idoxuridine. Observation was continued for 14 days and recurrence of dendritic ulceration took place in nine eyes. No curative effect was seen in the 16 eyes treated with the placebo solution.

During the course of studies on the inhibitory effect of idoxuridine against herpes virus in tissue culture in a liquid medium Kaufman and Maloney (1963 b) observed that the activity of solutions decreased during storage. They attributed this to decomposition with formation of inhibitory products, and they claimed that the antiviral action of 1 mg/ml idoxuridine could be reversed by the breakdown product 5-iodouracil in concentrations as low as 0·1 µg/ml. This effect could not be confirmed by Engle and Stewart (1964), who tested 5-iodouracil in parallel with thymidine, a known inhibitor of idoxuridine. In plaque inhibition tests the plaque-free zone produced by a disc containing 50 µg idoxuridine was abolished when a disc containing 10 µg of thymidine was placed in close proximity, but was unaffected by a disc containing 100 µg 5-iodouracil. It is difficult to reconcile this discrepancy, apart from the fact that a different method of testing was used.

Idoxuridine was also found to be effective in the treatment of experimental herpes infection of the skin in guinea-pigs (Tomlinson and Mac-Callum, 1968). The hair was removed from the backs of the animals and a suspension of herpes virus was inoculated by multiple pressures from a vaccination needle. Vesicles appeared after 3 days and reached maximum development on the 6th day of infection. The development of the lesions could be entirely prevented by painting the affected area with a 9% solution of idoxuridine in 90% aqueous dimethylsulphoxide three times daily for 3 days, beginning 24 hours after infection.

Early work on the chemotherapy of herpes was carried out with untyped strains, but it was evident that they varied considerably in their sensitivity to idoxuridine. Thus de Lavergne *et al.* (1965) found that the minimum concentration which completely inhibited the growth of herpes virus in monkey kidney cells ranged from 30 µg/ml or less to 500 µg/ml. Person *et al.* (1970) compared 21 strains of known type in plaque inhibition tests. In chick embryo fibroblasts type 1 strains were more sensitive than type 2, but there was no marked difference in sensitivity when WI-38 or HeLa cells were used. There was some evidence that the sensitivity of type 2 strains could be increased by serial passage in HeLa cells, and it is evident that passage history and cell type should be taken into account when assessing the chemotherapeutic sensitivity and resistance of strains of herpes virus. Lowry *et al.* (1971) determined sensitivities in rabbit kidney cell monolayers by two methods. Virus preparations were titrated by plaque counting on monolayers in the presence of 25 µg/ml idoxuridine in parallel with a similar titration in which idoxuridine was not present. A 230-fold decrease in titre was found in titrations of 25 type 1 strains, and only a 2-fold reduction with 40 type 2 strains. Plaque reduction tests were also carried out with varying concentrations of idoxuridine in the overlay. The method gave 70% reduction concentrations of 2·0–8·0 µg/ml for type 1 and 28·8–43·8 µg/ml for type 2.

Lerner and Bailey (1972) determined 50% inhibitory concentrations in BHK-21 monolayers under overlay. If their results are adjusted to make them comparable with those of other workers, it is found that the inhibitory concentration for 11 type 1 strains ranged from 0·625 µg/ml to 2·5 µg/ml, and for three type 2 strains it was 12·5 µg/ml. Fiala *et al.* (1974) found that the minimum concentrations which completely inhibited the cytopathic effect of around 100 TCID50 of herpes virus in human embryo lung fibroblasts ranged from 37 µg/ml to 75 µg/ml for 11 type 1 strains and from 25 µg/ml to 75 µg/ml for 15 type 2 strains. They could therefore find no difference in type sensitivity.

One is left with the impression that type 2 strains may be somewhat less sensitive than type 1, but there is no general agreement on the point. It seems likely that the discrepant results obtained by different methods will be of little clinical relevance, since sensitivity evidently depends upon the type of cell in which the virus is growing, and this will not be the same as that involved in the natural infection.

Buthala (1964) observed that when the multiplication of herpes virus in monolayers of RK cells was inhibited by idoxuridine in concentrations up to 1000 µg/ml the virus infection would eventually break through and destroy the culture if incubation was continued. This was not due to decomposition of the compound, since it still occurred if the idoxuridine was replaced each day. In similar tests carried out under a solid overlay some plaques could always be obtained, and virus isolated from these plaques was found to be resistant to idoxuridine, since it would multiply to a normal extent in the presence of idoxuridine in a concentration of 500 µg/ml. This resistant strain retained its sensitivity to cytarabine. It was stable, since it retained its resistance during serial passage in medium not containing idoxuridine. Herpes virus induces the formation of thymidine kinase in infected cells, and resistance to idoxuridine might be due to the formation of a mutant virus which lacks the ability to code for this enzyme but can multiply by using the thymidine kinase of the cell. However, Centifanto and Kaufman (1965) found that a resistant strain grown in L cells caused the incorporation of [^{3}H]thymidine into DNA at the same rate as a sensitive strain, and a mutant strain of herpes virus which was unable to induce thymidine kinase was still sensitive to idoxuridine. Resistance must therefore be due to some other underlying mechanism, which has so far not been elucidated. Herpetic keratitis in man may appear to be resistant to treatment with idoxuridine; Coleman et al. (1968) found that 12 strains isolated from such cases were mostly sensitive to idoxuridine. Resistance to treatment is due to other factors, and there is no evidence that widespread use of idoxuridine has led to the emergence of resistant strains of virus, although the effect has been demonstrated experimentally in herpetic keratitis in rabbits (Underwood et al., 1965).

Idoxuridine also inhibits the growth of other members of the herpesvirus group. The activity against pseudorabies virus in tissue culture has been described above (Hermodson and Dinter, 1962). In animal experiments, however, the effect is only marginal. Kolb et al. (1963) infected the scarified corneas of rabbits with pseudorabies virus and treated the eyes with drops of 0·1% and 0·6% idoxuridine solution given at intervals of 30–120 minutes. Symptoms of infection appeared after 54·6–62·0 hours, compared with 35·8 hours in an untreated control group. A similar result was obtained by Benda (1965) in rabbits infected on the scarified cornea with B virus, a herpes virus of monkeys transmissible to man. Keratoconjunctivitis developed, and further progress of the infection led to death. The progress of the lesions could be retarded by instillations of 0·1% idoxuridine solution, or 0·2% ointment, but there was no reduction in mortality. Activity against B virus was later demonstrated in tissue culture by Miller (1967). In plaque reduction assays carried out in monkey kidney and HEp-2 cells complete inhibition of plaque formation was obtained with idoxuridine present in the overlay in a concentration of 78 µg/ml, and in plaque inhibition tests a zone of protection was obtained with a disc containing 400 µg, but not with 125 µg. These concentrations and amounts are rather high, and it is evident that B virus is not so sensitive to idoxuridine as herpes.

Rawls et al. (1964) found that the formation of plaques in tube cultures of WI-38 cells by varicella-zoster virus was completely suppressed when idoxuridine was present in the overlay medium in a concentration of 5 µg/ml, and they therefore concluded that varicella was a DNA virus, a fact which was not known at that time. The antiviral effect was confirmed by Rapp (1964), who also noted that idoxuridine was less active against varicella-zoster than cytarabine.

De Lavergne *et al.* (1965) found that the growth of cytomegalovirus in human diploid cells was inhibited by idoxuridine, although concentrations as high as 100 µg/ml were required. The facilitation of the growth of cytomegalovirus produced by the pretreatment of cells with idoxuridine has already been described (St. Jeor and Rapp, 1973).

The activity of idoxuridine against vaccinia was first detected by Herrmann (1961), using the newly developed plaque inhibition test. Calabresi *et al.* (1962) reported that antiviral activity could also be demonstrated in rabbits. The animals were inoculated intradermally with serial decimal dilutions of vaccinia virus, and idoxuridine was administered subcutaneously every 4 hours for 5 days in doses of 60 mg/kg. The development of skin lesions was markedly suppressed in comparison with untreated control animals. Kaufman *et al.* (1962 a) showed that idoxuridine was effective in curing vaccinial keratitis in rabbits. The corneas were scarified and infected with vaccinia virus, and treatment was begun 48 hours later, when corneal ulceration was well established. Idoxuridine was instilled 2-hourly for 2–4 days as a 0·1% solution. After 48 hours of treatment the lesions had disappeared, whereas the infection was still progressing in untreated control animals. Fulginiti *et al.* (1965) were unable to confirm this effect. The reason for this is not apparent, since the curative effect of idoxuridine in vaccinial keratitis can be demonstrated very easily, as will be shown when the treatment of vaccinial keratitis in man is discussed later.

Vaccinia virus can be made resistant to idoxuridine. Loddo *et al.* (1963) passaged vaccinia virus in a continuous line of human amnion cells in the presence of increasing concentrations of idoxuridine. Resistance developed and the virus was eventually able to multiply in the presence of idoxuridine in concentrations of 2000 µg/ml. Resistant strains of vaccinia have not been encountered in clinical practice. This is undoubtedly due to the fact that vaccinia infections arise mainly from vaccination against smallpox, and only to a very limited extent from case to case transfer.

Idoxuridine also inhibits the multiplication of fibroma and myxoma viruses, two further members of the poxvirus group. A concentration of 10 µg/ml in the overlay gave 100% reduction in plaque formation in RK 13 cells with both viruses (Roby *et al.*, 1965).

There is little to indicate that idoxuridine has any effect against adenoviruses. De Lavergne *et al.* (1965) found that the multiplication of types 1, 2, 3 and 9 in monkey kidney calls was not inhibited by concentrations of idoxuridine as high as 1000 µg/ml, although type 5 was inhibited by 500 µg/ml. Herrmann (1968) reported that types 1, 2, 5 and 7 were not inhibited by idoxuridine in plaque inhibition tests. Bauer and Apostolov (1966) found that idoxuridine produced a dose-related inhibition of the production of adenovirus type 11 haemagglutinin in tube tests, although complete inhibition was not obtained at the highest concentration tested. The production of infective SV15 virus (a simian adenovirus) in tissue culture was inhibited by idoxuridine in concentrations of 50 μg/ml and 100 μg/ml, but the production of haemagglutinin was not affected (Omelchenko *et al.*, 1969). It seems therefore that idoxuridine may have some effect against adenoviruses, but that the activity is not great enough to be of clinical interest.

Idoxuridine is also active against certain members of the papovavirus group. Munyon *et al.* (1964) found that the multiplication of polyoma virus in mouse embryo cells was inhibited by a concentration of 10 μg/ml (28 μM) when the compound was added to the cultures between 4 and 22 hours after infection.

It will also inhibit SV40, a papovavirus of monkeys (Haas and Maass, 1964), and Manilla *et al.* (1965) claimed that human warts could be cleared by repeated applications of a mixture of equal parts of 0·5% idoxuridine solution and hydrophilic cream base. This observation cannot be confirmed by laboratory work since wart virus has not yet been grown in existing tissue culture systems.

The antiviral spectrum of idoxuridine is shown in Table 3.4. It is used clinically in the treatment of ocular and cutaneous herpes and zoster

Table 3.4   Reported spectrum of antiviral activity of idoxuridine

Type	Group	Virus
DNA	Herpesvirus	Herpes 1
		Herpes 2
		B virus
		Pseudorabies
		Malignant catarrh
		Bovine rhinotracheitis
		Varicella-zoster
		Cytomegalovirus
	Poxvirus	Vaccinia
		Fibroma
		Myxoma
	Adenovirus	Adenovirus 5
		Adenovirus 11
		SV 15
	Papovavirus	Polyoma
		SV 40
		Verruca
	Iridovirus	African swine fever
RNA	None	None

infections, and with somewhat uncertain results in the treatment of herpetic encephalitis and congenital cytomegalovirus infections. On account of the toxicity and rapid metabolic degradation of idoxuridine there is a tendency now to use trifluorothymidine or vidarabine in its place.

Toxicity

Studies on the antiviral effect of idoxuridine in tissue culture have yielded incidental information on its toxicity. Resting cells are relatively unaffected, but the compound is toxic to dividing cells. Munyon *et al.* (1964) found that a concentration of 10 µg/ml prevented more than one division of mouse embryo cells, and that the inhibition could not be reversed by washing and treatment with thymidine. In similar experiments Buthala (1964) observed that the multiplication of rabbit kidney cells was inhibited irreversibly by concentrations of 12·5 µg/ml and higher, whereas de Lavergne *et al.* (1965), in tests apparently carried out with formed monolayers, found that concentrations up to 300 µg/ml were well tolerated by a range of cell types. The plating efficiency of HeLa cells, which reflects their ability to divide, was reduced to 20% by 10 µM idoxuridine and to zero by 1000 µM; the concentration giving 50% reduction was 1 µM (Umeda and Heidelberger, 1969).

Prusoff *et al.* (1960) found that the acute LD50 in mice on intraperitoneal injection was 2·5 g/kg, and the 13-day LD50 was 318 mg/kg. The signs of toxicity could be completely reversed by the administration of thymidine.

A great amount of information on the toxicity in man has become available from early studies on the use of idoxuridine in the treatment of malignant conditions, and subsequently in the treatment of herpetic encephalitis. In these conditions idoxuridine is given by intravenous infusion. Calabresi (1963) observed the development of leukopenia, thrombocytopenia, stomatitis, alopecia and transverse ridging of the nails, which were presumably due to the inhibitory effect of idoxuridine on cell division. They appeared in all patients who received a total dose of more than 600 mg/kg and in two-thirds of those given 400–600 mg/kg. These signs of toxicity were not observed with doses less than 400 mg/kg. Other manifestations included anorexia, less frequently nausea and vomiting, and occasionally mild signs of iodism, with oedema of mucous membranes, ptyalism and acneform dermatitis (Calabresi *et al.*, 1961). Dayan and Lewis (1969) observed the appearance of jaundice with abnormal liver function tests in an adult patient with herpetic encephalitis treated with 30 g of idoxuridine given in divided doses over 8 days. The patient subsequently died, and at autopsy the liver was deeply jaundiced; there were no inflammatory or necrotic areas, but there were many bile thrombi. Idoxuridine can thus produce cholestatic jaundice. Silk and Roome (1970) observed elevation of serum bilirubin and alkaline phosphatase in a 7-year old boy with herpetic encephalitis who received 11·5 g over 8 days, and they stated that liver damage should be expected in patients treated with idoxuridine. Nolan *et al.* (1973) have analysed the toxicity data from 29 patients with herpetic encephalitis treated with idoxuridine. The maximum effect with minimum toxicity was obtained with 54 mg/kg daily for 5 days given as two rapid infusions of 50 mg per minute. A total adult dose of 20 g should not be exceeded, or thrombocytopenia will occur, leading to haemorrhages in the nasal mucosa, bowel, trachea and urinary tract. Bacterial infections may be difficult to recognize, as the resultant leuko-penia reduces the extent of purulent exudates.

Signs of toxicity are not infrequently observed when solutions of idoxuri-dine are applied to the eye. Corrigan *et al.* (1962) used a 0·1% solution for the treatment of 15 patients suffering from herpetic keratitis and observed toxic effects in six patients. In two, secondary non-dendritic ulceration and epithelial degeneration appeared at the site of the original ulcer. In these and four other patients there was considerable punctate keratitis after treatment for 5–6 days. These changes disappeared when treatment was stopped, except in one patient in whom a large erosion was still present after 20 days. These effects were assumed to be due to the inhibitory effect of idoxuridine on DNA metabolism in the nuclei of the epithelial cells.

Patterson and Jones (1967) observed signs of toxicity in seven of 50 patients with ocular herpes who had been treated with idoxuridine eye drops for periods ranging from 39 to 112 days. Follicular conjunctivitis and oedema developed, leading to lacrimation and heaviness of the lid. If treatment was continued the lacrimal puncta became occluded in 1–2 weeks. The signs disappeared rapidly when treatment was discontinued. No changes were observed in the cornea or lens. In two patients an allergic reaction was seen, consisting of contact dermatitis and follicular conjuncti-vitis. It was established that the other components of the eye drops were not responsible for the allergic reaction.

Idoxuridine given after keratoplasty has been reported to be toxic to the epithelium, which may become macerated. Also, the epithelium over the graft appears to be more sensitive to the toxic effects of idoxuridine than the host epithelium (Rice and Jones, 1973).

Metabolism

Early studies of the pharmacology of idoxuridine in patients with advanced malignant conditions indicated that it was rapidly metabolized. Thus, after the administration of labelled idoxuridine by intravenous drip in a total dose of 80 mg/kg, 40% of the radioactivity was present in the urine after 4 hours, and was in the form of 5-iodouracil, which has no antiviral activity, and iodide (Welch *et al.*, 1960). Similar results were obtained by Prusoff *et al.* (1960) in mice injected by the intraperitoneal route.

The metabolism of idoxuridine in the brain and cerebrospinal fluid was studied by Clarkson *et al.* (1967). In dogs injected intravenously with labelled compound only very low amounts of radioactivity appeared in the cerebrospinal fluid, and 80% of this was accounted for by iodide ion. Idoxuridine disappeared rapidly from the plasma, with the appearance of iodouracil and iodide ion. When idoxuridine labelled both with ^{125}I and ^{3}H was injected into the cisterna magna ^{125}I disappeared twice as rapidly as ^{3}H. This implies that the compound was metabolized rapidly with the liberation of iodide ion, and this was confirmed by the isolation of metabolites from the cerebrospinal fluid, when it was found that idoxuridine was rapidly converted into iodouracil. There was no breakdown when idoxuridine was incubated with cerebrospinal fluid *in vitro*, but it was rapidly metabolized when incubated with brain slices and microsomal preparations. These results indicate that idoxuridine fails to enter the central nervous system in significant amounts, and may explain the general failure of attempts to use idoxuridine in the treatment of herpetic encephalitis.

Clinical pharmacology

Prusoff *et al.* (1960) injected labelled idoxuridine intravenously into mice and found that the radioactivity was generally distributed in the internal organs. The concentrations fell rapidly by 2 hours, and the compound had largely disappeared by 24 hours. A considerable amount of the radioactivity was found in nucleic acids and protein mainly as a result of the incorporation of iodide ion. Welch *et al.* (1960) studied the fate of labelled idoxuridine in patients with advanced malignant conditions. Four hours after the end of administration in an intravenous drip 43% of the radioactivity had passed into the urine, and of this 60% was mainly 5-iodouracil and the remainder inorganic iodide; 89% of the radioactivity was excreted in 24 hours and 95% in 48 hours. Some radioactivity was found in the nucleic acid fraction of a tumour biopsy taken 70 hours after administration. As described already in the section on metabolism, Clarkson *et al.* (1967) found that idoxuridine was unable to enter the central nervous system of dogs after intravenous injection. It presumably cannot cross the blood-brain barrier, but it is conceivable that this might not be functionally intact in herpetic encephalitis, thus permitting some entry to the cerebrospinal fluid, although in concentrations too low to have any definite clinical effect.

Lerner and Bailey (1972) determined levels of idoxuridine in nine patients by means of a microbiological method. It was generally given by continuous intravenous infusion for 5 days to a total dose of 300–400 mg/kg at a rate of 4 mg per minute. In 22 samples of serum taken during the course of treatment idoxuridine was detected in only two instances, in concentrations of 33 and 9 µg/ml. Six of 10 samples of urine contained amounts ranging from 45 to 80 µg/ml. Eight samples of cerebrospinal fluid were examined; idoxuridine was found in only one, in a concentration of 1040 µg/ml, but this sample had been obtained immediately after the beginning of transfusion. One patient received a rapid infusion of idoxuridine given at the rate of 50 mg per minute. Serum samples collected from 15 to 60 minutes after beginning the transfusion contained amounts of idoxuridine ranging from 10 to 36 µg/ml, and a sample of urine obtained after 45 minutes contained 1040 µg/ml.

The behaviour of idoxuridine in the eye has been studied in experiments carried out in rabbits (Mastan and Henderson, 1966). A 0·2% solution was injected by the transdermal subconjunctival route in one eye, and the concentrations in various tissues and media were determined at intervals. A peak concentration of 12·6 µg/g was attained in the cornea after 1 hour. Other peak values were 4·8 µg/g in the iris after 15 minutes and 2·6 µg/g in the aqueous after 2–3 hours. A therapeutic level of idoxuridine in the cornea was attained after 30 minutes and persisted for 4 hours.

Bakhle *et al.* (1965) investigated the incorporation of labelled idoxuridine during chronic administration. Two drops of 0·1% or 0·3% solution were placed in the conjunctival sac of rabbits every 30 minutes for 8 hours on 3 successive days. Two days after the last instillation the distribution of the radioactivity was determined in various fractions prepared from the tissues. There was significant incorporation in the corneal epithelium and conjunctiva, and lesser uptake into the lens and iris. Most of the radioactivity was present in the cold acid-soluble fraction, which indicated that the compound was present in an uncombined form, but there was some incorporation into nucleic acid in the corneal epithelium. Significant amounts of radioactivity were found in the thyroid and bone marrow, and lesser amounts in most internal organs. This was attributed to passage through the nasolacrimal duct and absorption from the gastrointestinal tract. The uptake in the thyroid was presumably due to degradation to iodide. The authors concluded that systemic toxicity is unlikely to result from the application of idoxuridine to the eye, since the amounts present in the organs are only 1/100th of the toxic level.

Assay methods

The method developed by Buthala (1964) for the assay of cytarabine can also be used for the determination of idoxuridine in biological fluids, and will not be repeated here. Levels of less than 1 µg/ml can be detected in tissue extracts, and 2 µg/ml can be detected in serum samples which are first treated with acetone to remove herpes antibody.

A similar method has been used by Lerner and Bailey (1972), which is also based on the activity against herpes virus. Monolayers of BHK-21 cells in 30 ml flasks are infected with 50–100 pfu of herpes virus in a volume of 0·2 ml. After adsorption for 20 minutes the cultures are washed and

0·4 ml volumes of 2-fold dilutions of the test samples are added; 4 ml of agarose overlay medium are poured on the cultures and allowed to set, and incubation is continued for 96 hours at 35 °C. The cultures are then fixed with formalin and stained. The number of plaques is counted, and the amounts of idoxuridine in the test samples are obtained from a standard curve relating plaque count to the concentration of the compound. The sensitivity of the method is around 2·5 µg/ml.

Clinical use

Idoxuridine is used for the treatment of infections caused by viruses of the herpes group. The indications are shown in Table 3.5.

Table 3.5   Clinical use of idoxuridine

Group	Virus	Condition	Route	Preparation
Herpesvirus	Herpes	Keratitis	Topical	Eye drops
		Cutaneous herpes	Topical	Solution in DMSO
		Whitlow	Topical	Solution in DMSO
		Encephalitis	Intravenous	Aqueous solution
	Varicella-zoster	Zoster	Topical	Solution in DMSO
	Cytomegalo-virus	Cytomegalic inclusion disease	Intravenous	Aqueous solution

Contraindications

Treatment of herpetic keratitis with idoxuridine should be discontinued if signs of toxicity develop. It should not be given systemically during pregnancy as it has been shown to be a potent teratogenic agent in mice.

Preparations

Idoxuridine eye-drops (BNF). Sterile 0·1% aqueous solution. The container should be discarded not later than 4 weeks after opening. Idoxuridine ophthalmic solution (USP). Sterile 0·1% aqueous solution which may contain buffers, stabilizers and antibacterial agents. The pH may range from 5 to 7·5.
Dendrid (Alcon, USA). Smith and Nephew Pharmaceuticals. Eye dropper containing 0·1% idoxuridine in 15 ml drop per bottles.

Kerecid (Stoxil, USA). Smith, Kline and French. Eye dropper containing 0·1% idoxuridine and 0·002% thiomersal in 15 ml drop per bottles.

Herpid. W. B. Pharmaceuticals. A solution containing 0·5% idoxuridine in dimethylsulphoxide.

Solutions of idoxuridine should be stored in a cool place and protected from light.

Idoxuridine opthalmic ointment (USP). Contains 0·5% idoxuridine in soft paraffin base.

TRIFLUOROTHYMIDINE

Trifluorothymidine is the generic name of 5-trifluoromethyl-2'-deoxyuridine (Figure 3.3). It is often referred to in the literature as F_3TdR (trifluorothymine deoxyriboside). It is a synthetic nucleoside containing the naturally occurring sugar deoxyribose, and was synthesized by Heidelberger *et al.* (1964 b) during the course of studies of the biological effects of synthetic pyrimidine derivatives. Its solubility in water does not appear to have been reported, but Sugar *et al.* (1973) used a 5% solution in experimental work, and stated that one of the advantages of trifluorothymidine was its high solubility.

Aqueous solutions of trifluorothymidine are rather unstable. The optimum pH range for stability is 1–4. At pH 7·4 the half-life is 1·5 day at 30 °C and 16 hours at 37 °C. The products of degradation are fluoride ion, 5-carboxymethyl-2'-deoxyuridine and 5-hydroxydifluoromethyl-2'-deoxyuridine. Trifluorothymidine has been mainly used on a research basis up till now, and solutions for ophthalmic treatment have been made up at the time of use and stored in the refrigerator for a few days only. It is not given by intravenous injection.

Trifluorothymidine is very soluble in dimethylsulphoxide, but there have been no reports of the clinical use of such solutions.

Biological actions

The biological actions of trifluorothymidine reflect its involvement in the pathways of DNA synthesis. Gottschling and Heidelberger (1963) found that a concentration of 200 µg/ml would produce a 30-fold increase in the mutation rate of the T4 phage. From 10 to 15% of the DNA extracted from the phage banded at a higher density in caesium chloride density gradients, an observation which indicated that trifluorothymidine had been incorporated to some extent. Trifluorothymidine is also incorporated into virus DNA, and this will be described in a later section.

The possibility of mutagenic action was also investigated by Huberman and Heidelberger (1972), who used the method of development of resistance to the inhibitory effect of 8-azaguanine in Chinese hamster cells exposed to various concentrations of the compound for 4 hours and then grown in the presence of the inhibitor for 48 hours. Growth of resistant colonies was not noted, and this indicated that trifluorothymidine had not caused mutation. In similar experiments cytarabine was found to be highly mutagenic.

The compound inhibits cell division. Szybalski *et al.* (1963) found that it was highly inhibitory to the human cell line D985, a concentration of 0·012 µg/ml reducing the plating efficiency (ability to divide) by 50%. This inhibition was reversed if thymidine was added in a concentration slightly less than equimolar. Incorporation into cell DNA was observed in density gradient experiments.

An antitumour effect has been observed both in animals and in man. Heidelberger and Anderson (1964) found that trifluorothymidine inhibited the growth of the L1210 leukaemia and adenocarcinoma 755 cell lines. Ansfield and Ramirez (1971) gave intravenous injections of trifluorothymidine to 43 patients with disseminated cancer; a reduction in tumour size of 50% or more occurred in eight of 23 patients with breast cancer, and there was almost complete remission of carcinoma of the colon in one of six patients, but the condition began to progress again after the end of treatment. Natarajan and Rao (1965) studied the effect of trifluorothymidine on chromosomes in preparations of rootlets of *Vicia faba* which had been exposed to a 1 mM solution for up to 10 hours. Chromosome preparations showed Feulgen-negative achromatic regions, chromatid gaps, chromosome gaps, chromatid breaks and isolocus breaks. Chromosome breaks could be found after exposure for only 1 hour. These results indicate that trifluorothymidine is a powerful chromosome-breaking agent, but only in rather high concentration. It is therefore not surprising that trifluorothymidine has been found to be teratogenic (Kury and Crosby, 1967). When injected in a range of doses into the yolk sac of fertile eggs between the 34th and 90th hours of incubation it was found to be consistently teratogenic. The effective dose range for teratogenicity was 0·1–0·5 µg when given on the second day of incubation, 0·75–1·75 µg on day 3, and 2·4–3·1 µg on day 4. The abnormalities mainly affected the skeletal system, with cleft palate, absence of limbs, shortening or curvature of individual bones. Bilateral hypoplasia of the kidneys was occasionally observed in 19–20 day embryos, which was probably due to massive renal haemorrhage. Necrosis of limb, head and trunk mesenchyme was seen in earlier embryos removed 3 days after injection.

Mode of action

The substitution of three fluorine atoms into the 5-methyl group of thymidine to form trifluorothymidine leads to changes in the properties of the molecule which are presumably the basis of its mode of action. The effect is to increase the acidity of the proton in the 3-position, so that the pK_a for 5-trifluoromethyluracil (the pyrimidine moiety of the molecule), falls to 7·35 from the value of 9·82 characteristic of thymine (Heidelberger *et al.*, 1964). This change in properties would presumably affect the base-pairing mechanism of DNA, and also the interaction of trifluorothymidine with enzymes.

Trifluorothymidine has several biochemical actions. Gottschling and Heidelberger (1963) found that it inhibited the thymidylate synthetase activity of extracts of Ehrlich ascites carcinoma cells. This observation was confirmed by Heidelberger *et al.* (1964 a) in suspensions of the same tumour cells. Trifluorothymidine inhibited the incorporation of labelled formate into DNA thymine as a result of inhibition of thymidylate synthetase. An inhibition of 70% was produced by a concentration as low as 0·6 µM. The active species was shown to be the 5'-monophosphate of trifluorothymidine, which is produced after entry into the cell. This was confirmed by Reyes and Heidelberger (1965), who found that the 5'-mono-phosphate would inhibit a preparation of thymidylate synthetase from Ehrlich carcinoma cells, and slowly inactivate the enzyme irreversibly. Inhibition could also be demonstrated *in vivo* in mice bearing the tumour and given a single intraperitoneal dose of 200 mg/kg of the unphosphory-lated compound. The phosphorylation of trifluorothymidine was studied by Bresnick and Williams (1967), who found that a preparation of deoxy-thymidine kinase from regenerating rat liver phosphorylated it as readily as the natural substrate deoxythymidine. Trifluorothymidine also inhibited the enzyme, and thus the formation of deoxythymidine monophosphate from deoxythymidine.

Trifluorothymidine thus blocks stages in the thymine pathway, and the overall effect on DNA synthesis was studied by Fujiwara *et al.* (1970). When HeLa cells were incubated with a 1 µM solution for 48 hours no division took place and there was considerable death of the cells. Labelled trifluorothymidine was incorporated into cell DNA to the extent of 12% of the normal content of thymidine. DNA could thus be synthesized in the presence of trifluorothymidine, but cell division was prevented. On density gradient analysis the DNA was found to be in the form of short strands, corresponding to one or two breaks in the normal DNA strand.

The multiplication of vaccinia virus is inhibited by trifluorothymidine, as will be described in the following section, and further information on the mode of action of the compound has been obtained in studies of cells infected with vaccinia virus in the presence of trifluorothymidine. Umeda and Heidelberger (1969) found that the multiplication of the virus was completely arrested by a concentration of 0·1 µM, and the compound was not toxic at this concentration, as determined by the effect on plating efficiency. There was no recommencement of virus multiplication after the compound had been washed off, and the antiviral effect was thus irreversible. It could, however, be reversed if 1 μM thymidine was added to the cultures simultaneously.

Fujiwara and Heidelberger (1970) found that the DNA of vaccinia virus grown in HeLa cells in the presence of 1 μM labelled trifluorothymidine contained the compound to the extent of 1·4–9·8% of the normal content of thymidine. The cells contained only 6% of the infectivity of control cultures not exposed to trifluorothymidine, and this could have represented the residue of the input virus to a certain extent. Density gradient studies showed that the DNA of normal virions was 70S, whereas DNA from virions containing trifluorothymidine was 52S and 32S, which indicated that breakage of the strands had occurred. Uncoating of the virus took place normally in cultures exposed to trifluorothymidine, but only 50% of the virus DNA formed was finally encapsulated. On electron microscope examination the virions were circular in section, and had thick walls with little internal structure.

In further studies carried out in HeLa cells in the presence of 1 μM trifluorothymidine it was found that early mRNA was formed and associated normally with ribosomes, but there was no association of late mRNA with ribosomes, presumably because none was formed. This was interpreted as the result of incorporation of trifluorothymidine into the progeny DNA, which would render it unable to transcribe late mRNA, whereas the early mRNA, transcribed by the parental genome, could be formed normally. This supposition was supported by the further observation that when cells were infected with virus containing trifluorothymidine in its DNA the early mRNA was not transcribed normally. The results of polyacrylamide gel electrophoresis indicated that a major virus-induced protein was not synthesized in the presence of trifluorothymidine, and that another protein was formed instead (Oki and Heidelberger, 1971).

On the basis of the above evidence it seems likely that the antiviral action of trifluorothymidine results from its incorporation into virus progeny DNA, which thus disturbs the late transcription functions, leading to an abnormality of protein synthesis which affects encapsulation.

Antiviral activity

The antiviral activity of trifluorothymidine was apparently first observed in animal experiments and the literature on the subject is very sparse in comparison with other antiviral compounds. Kaufman and Heidelberger (1964) infected rabbits with herpes virus on the scarified cornea, and 3 days later when ulcers had developed, the eyes were treated with drops of trifluorothymidine or a placebo every 2 hours for 2 days. The extent of the ulcers was then scored by a numerical method. Trifluorothymidine given in a concentration of 0·1% was effective in eliminating the ulcers. It was equally effective in the treatment of ulcers produced by a strain of herpes virus which had been made resistant to idoxuridine. Shen *et al.* (1966) stated that trifluorothymidine inhibited herpes, vaccinia and type 2 adeno-virus (presumably in tissue culture), but no experimental details were given. Hyndiuk *et al.* (1968) found that it was active against herpes virus in plaque inhibition tests; a disc containing 10 μg gave an extensive zone of protection without any signs of toxicity.

Sugar *et al.* (1974) studied the effect of trifluorothymidine on herpetic iritis. A strain of herpes virus which produced iritis was injected into the anterior chamber of rabbits. Treatment with 5% trifluorothymidine drops given every 2 hours reduced the development of iritis in comparison with untreated control eyes.

The effect of solutions of trifluorothymidine in dimethylsulphoxide on cutaneous herpes in guinea-pigs was studied by Bauer and Collins (1974). The animals were shaved and herpes virus was inoculated by scarification at eight sites on both flanks. After 24 hours, when lesions were developing, the sites of infection on one flank were painted twice daily with trifluoro-thymidine solution and the lesions on the other flank were painted with dimethylsulphoxide alone for control purposes. After 4 days the lesions treated with trifluorothymidine had healed, whereas the lesions on the control side showed erythema, induration, vesiculation and pustulation, and healing was not complete until 7 days later. Satisfactory healing was obtained with solutions of 5% or over; some effect could still be detected with a 1% solution, but the rate of healing was not rapid enough to be clinically acceptable.

Trifluorothymidine inhibited the multiplication of vaccinia virus in suspended cultures of HeLa cells in concentrations as low as 0·1 μM, and was not toxic at this level (Umeda and Heidelberger, 1969). Trifluoro-thymidine has some effect in experimental vaccinial keratitis in rabbits, but the activity is not great enough to justify its use in clinical practice in view of the greater activity of cytarabine and idoxuridine (Bauer and Collins, 1974).

There are as yet no reports of activity against other members of the herpes and pox virus groups, and the antiviral spectrum of trifluorothymi-dine is thus limited to herpes, vaccinia and adenovirus.

Toxicity

The short-term toxicity of trifluorothymidine was determined by Heidel-berger and Anderson (1964). In mice given daily intraperitoneal injections for 7 days the LD50 was 325 mg/kg. It was more toxic when given by stomach tube, with an LD50 of 225 mg/kg. Palm *et al.* (1967) investigated toxicity in dogs and monkeys. During daily intravenous administration for 14 days leukopenia and occasional hypoglycaemia were observed with dose levels of 12·5 mg/kg or higher. The serum calcium and phosphate were slightly depressed with 25 mg/kg, and dose levels of 100 mg/kg caused leukopenia, transitory fall in the red cell count, decreased haematrocrit and haemoglobin, and elevated serum glutamic–pyruvic transaminase and blood urea nitrogen levels. The serum calcium and phosphate were slightly depressed. Higher doses caused hypoplasia of the bone marrow and focal degeneration of the proximal tubules of the kidneys and liver parenchyma. Intestinal haemorrhages occurred in a few animals. Helson *et al.* (1970) investigated the toxicity of trifluorothymidine in 18 children and 24 adults with advanced neoplastic disease. The children tolerated 30 mg/kg daily or every other day for 8–12 doses, to a total dose of 105–880 mg/kg, with a mean of 350 mg/kg. These doses produced leukopenia, reticulocytopenia, anaemia, megaloblastosis and mild hypocalcaemia. Drug fever occurred in three children. The adults received only 5 mg/kg for six doses. The signs of toxicity were haematological also, with a high incidence of mild hypocalcaemia and hypophosphataemia. Gastrointestinal toxicity was not encountered, although four children developed stomatitis. Similar studies were carried out by Ansfield and Ramirez (1971) in 43 patients with disseminated cancer. Trifluorothymidine was given intravenously in five daily doses of 1·5–30 mg/kg by rapid injection of a 20 mg/ml solution in saline, or in a daily dose of 2·5 mg/kg given as divided doses for 8–13 days. There was moderately severe depression of the bone marrow. About one-half of the patients had white cell counts below 2000, and about one-third had platelet counts below 100 000. One-third had mild diarrhoea. Severe irreversible toxicity to the haematopoietic and gastrointestinal systems developed in one patient who was treated for 13 days; death ensued with severe vomiting, diarrhoea, leukopenia and thrombocytopenia.

These reports of toxicity after systemic administration are not contra-indications to the use of trifluorothymidine as an antiviral agent, since it is so far only used for the topical treatment of herpetic keratitis, and the amount absorbed into the body after administration by this route would be too small to cause any signs of general toxicity. Local toxicity in the eye has not yet been reported.

Metabolism

Heidelberger *et al.* (1963) found that a nucleoside phosphorylase from Ehrlich ascites carcinoma cells could remove the deoxyribose moiety from trifluorothymidine and thus convert it to trifluorothymine. Further work was carried out in mice bearing the ascites tumour and injected with [^{14}C]trifluorothymidine (Heidelberger *et al.*, 1965). Very little radioactivity appeared in the respiratory carbon dioxide and the pyrimidine ring was therefore not appreciably degraded. Trifluorothymidine appeared in the urine along with its breakdown products trifluorothymine and 5-carboxy-

2'-deoxyuridine. Trifluorothymidine accumulated in the liver and spleen and was incorporated into DNA, but only to the extent of 0·32%.

Dexter *et al.* (1972) carried out a similar investigation in eight patients with advanced cancer. Trifluorothymidine was given by intravenous injection or infusion in doses ranging from 0·3 to 27 mg/kg. The only metabolites found in serum and urine samples were trifluorothymine and 5-carboxyuracil. Unchanged compound was also present. Here again no labelled carbon dioxide could be detected. Measurable amounts of trifluorothymidine became bound to serum proteins. In investigations on eight children the compound was metabolized less rapidly than in adults.

A preparation of deoxythymine kinase from rat liver phosphorylated trifluorothymidine as readily as its natural substrate deoxythymidine (Bresnick and Williams, 1967). In the toxicological study in dogs and monkeys already described (Palm *et al.*, 1967), the total and inorganic fluorine content of the urine was determined. From the results it was inferred that 70% of the fluorine content of the trifluorothymidine injected may remain in the body, and that 5–10% of the body fluorine may be converted to 5-carboxyuracil and inorganic fluoride.

Clinical pharmacology

When labelled trifluorothymidine was given intravenously in monkeys and dogs the half-life was 30 minutes. By 24 hours 60–93% was excreted in the urine as unchanged compound, and the metabolites trifluorothymine, 5-carboxydeoxyuridine and 5-carboxyuracil, which have no antiviral activity. Monkeys converted trifluorothymidine to trifluorothymine 4–6 times as rapidly as dogs, and could therefore tolerate much higher doses. The tissues contained only 2–30 µg/g of trifluorothymidine equivalents (unchanged compound and its metabolites) after 24 hours. Trifluorothymidine was bound in the tissues, to the greatest extent in those tissues such as bone marrow and lymph nodes which showed the most signs of toxic damage. The binding was probably covalently linked and to protein (Rogers *et al.*, 1969). In the course of a study of the effect of 5% trifluorothymidine eye drops on herpetic iritis in rabbits, Sugar *et al.* (1973) studied its penetration into the aqueous. The concentrations were determined by a biological method described in the following section. In eyes infected with herpes virus the aqueous contained 37 µg/ml 30 minutes after administration, and 3·4 µg/ml after 60 minutes; none was present after 90 minutes. In uninfected eyes trifluorothymidine penetrated much less well, and levels of only 5–6 µg/ml were found.

Assay methods

A biological method of determining the concentration of trifluorothymidine in samples of aqueous has been briefly outlined by Sugar *et al.* (1973).

Samples were mixed with 900 pfu of a preparation of herpes virus and tissue culture medium to a volume of 1 ml; the mixtures were placed on monolayers of human embryo kidney cells and incubated for 3 days to allow plaques to develop. The number of plaques formed was reduced as a result of the antiviral effect of trifluorothymidine, and the reduction in comparison with infected control cultures not containing the compound was recorded and compared with that obtained with samples containing known amounts of trifluorothymidine.

An electrometric method has been developed by Rogers and Wilson (1969) for determining trifluorothymidine in samples of urine. The samples were made up to $0 \cdot 1$ N in sodium hydroxide and allowed to stand for 60 minutes at room temperature. Under these conditions hydrolysis takes place, and the fluoride ion which is liberated was determined with a fluoride ion electrode. The primary metabolite trifluorothymine is hydrolysed much more rapidly, and the difference in the rate of hydrolysis enabled both trifluorothymidine and its metabolite to be determined in the same sample. The sensitivity of the method is given as 5 µg/ml.

Clinical use

Trifluorothymidine is indicated for the treatment of herpetic infections of the eye. According to present opinion it is superior to idoxuridine, which gives rise to undesirable toxic reactions during prolonged treatment which may necessitate its withdrawal. Trifluorothymidine also has the advantage of being effective against strains of herpes virus which are resistant to idoxuridine. There have been no reports of the use of trifluorothymidine in the treatment of herpetic encephalitis, cutaneous herpes or zoster.

Contraindications

Contraindications to the use of trifluorothymidine have not yet become apparent. It has not been given intravenously in antiviral treatment, but if this route is selected for any reason its teratogenic potential would render it unsuitable for use during pregnancy.

Preparations

Trifluorothymidine (pure substance); Sas Scientific Chemicals Ltd.
There are no preparations available commercially.

VIDARABINE

Vidarabine (Figure 3.3) is the generic name of 1-β-D-arabinofuranosyladenine. It is commonly referred to in the literature as ara-A and adenine arabinoside. It is a nucleoside containing the uncommon pentose sugar arabinose in place of ribose or deoxyribose. It was originally synthesized as an antineoplastic agent by Lee *et al.* (1960), but has since been prepared by fermentation with the NRRL strain of *Streptomyces antibioticus.*

Vidarabine is a white crystalline powder with a maximum solubility in water of 0·05%. The maximum solubility in 0·1 M phosphate buffer at 37 °C is 0·18%. Solutions may be made up at this concentration and kept at 37 °C (Kurtz *et al.*, 1968). There is no information available on the stability of aqueous solutions, and it is therefore advisable to prepare solutions just before use.

Biological actions

The initial studies of vidarabine were devoted to a study of its potentialities as an antitumour agent. Brink and LePage (1964 a) found that it inhibited the growth of the TA3 and 6C3 HED ascites tumours in mice. Animals bearing implants of the tumour and treated with vidarabine in doses of 25 mg/kg given by an unstated route twice daily for 6 days survived for more than 50 days and appeared to be tumour-free, whereas untreated control animals died after 9–13 days. Further work (LePage and Junga, 1965) failed to confirm the activity against the TA3 tumour. However, it was found that mouse tissues contained adenosine deaminase which deaminated vidarabine. This effect could be blocked by the simultaneous presence of adenosine, which had a higher affinity for the enzyme, and some reduction in tumour growth was observed when the mice were treated with 40 mg/kg vidarabine combined with twice the weight of adenosine. Vidarabine has also been reported to inhibit cell growth in tissue culture. Doering *et al.* (1966) found that a concentration of 2×10^{-4} M inhibited the multiplication of mouse fibroblasts, but the activity of vidarabine in this respect was much lower than that of cytarabine, which produced the same effect in a concentration of 2×10^{-6} M. Shipman *et al.* (1973) found that the addition of 10–80 µg/ml vidarabine to a line of rat embryo cells transformed with Rous sarcoma virus growing in the early log phase brought about an inhibition of mitosis lasting for 5–30 hours, after which growth resumed at the normal rate. With a concentration of 160 µg/ml mitosis had not resumed after 30 hours and the number of visible cells had decreased. The inhibition could be reversed by removal of vidarabine up to 24 hours.

Vidarabine has some inhibitory activity against *Plasmodium berghei* (Ilan *et al.*, 1970). When mice infected with the parasite were treated with 0·5 g/kg daily the percentage of red cells becoming infected was significantly less than in untreated control animals, and the survival time was twice as long.

Mode of action

No studies of the mechanism of the antiviral action of vidarabine have appeared so far, and knowledge of its effects has been largely gained by the study of normal cell systems.

Hubert-Habart and Cohen (1962) found that the DNA content of *Escherichia coli* grown in the presence of 4 µg/ml vidarabine was reduced by 40%. The synthesis of RNA was unaffected, but its terminal nucleosides contained vidarabine to the extent of 10–20%, a substitution which might be expected to affect protein synthesis. Incorporation into RNA was also observed by Brink and LePage (1964) in ascites tumour cells; phosphorylation to the triphosphate took place before the incorporation stage. There was no incorporation into DNA. In further work on ascites tumour cells (York and LePage, 1966) vidarabine was found to be a non-competitive inhibitor of DNA polymerase, and this effect was considered to be the basis of its inhibitory action on cell growth. Inhibition of DNA polymerase was also observed by Furth and Cohen (1968), who worked with purified enzyme extracted from calf thymus and bovine lymphosarcoma cells. The same mechanism was considered to be responsible for the activity of vidarabine against *P. berghei* (Ilan *et al.*, 1970). However, the results of other studies are not in accordance with these findings.

Thus, Schnebli *et al.* (1967) found that adenosine kinase purified from HEp-2 cells did not phosphorylate vidarabine to any detectable extent. Lindberg *et al.* (1967) found that adenosine kinase prepared from rabbit liver and Ehrlich ascites tumour cells would phosphorylate vidarabine, but only to 5% of the extent observed with the natural substrate adenosine. LePage (1970) observed that vidarabine labelled in the ring system was incorporated into RNA in murine ascites turmour cells, and to a lesser extent into DNA, but this incorporation was found to be due to cleavage of the pentose moiety and reutilization of the free base. This mechanism might therefore explain the incorporation reported in earlier work. It may be concluded that phosphorylation of vidarabine is theoretically possible but does not occur to any significant extent under normal conditions.

Antiviral activity

The antiviral activity of vidarabine was first observed by De Rudder and Privat de Garilhe (1963), who found that a concentration of 330 µM reduced the yield of vaccinia virus in HeLa cells by 4 log units, and the yield of herpes virus by 5 log units. No further observations of antiviral activity were reported until 1968, when a number of papers appeared which have been summarized by Schabel (1968). In plaque reduction tests a concentration of vidarabine in the overlay of 31·2 µg/ml reduced to zero the production of plaques by vaccinia virus, and a concentration of 10 µg/ml gave 80% reduction. Vidarabine showed a similar degree of activity against herpes virus in the same test, and the virus did not become resistant during the course of three passages in the presence of 12·5 µg/ml vidarabine. There was also activity against cytomegalovirus, but to a lesser extent. This was confirmed by Sidwell *et al.* (1969), who found that vidarabine inhibited the growth of cytomegalovirus in WI-38 cells by 4 log units, and by Fiala *et al.* (1974), who found that the minimal concentration for 80% inhibition ranged from 6 to 200 µg/ml.

As is the case with other antiviral nucleosides, type 2 strains of herpes virus appear to be less sensitive to vidarabine than type 1 strains. This was demonstrated by Person *et al.* (1970) in plaque inhibition tests with several cell lines. Discs containing 100 µg of the compound gave inhibition zones of 26 mm mean diameter with 13 strains of type 1, and of only 3 mm with eight strains of type 2. A similar result was obtained by Fiala *et al.* (1974) by a different method. The minimal inhibitory concentrations of vidarabine ranged from 8 to 12 µg/ml with strains of type 1, and from 12 to 24 µg/ml with strains of type 2.

The yield of varicella virus in WI-38 cells was almost completely abolished by a concentration of 8 µg/ml. In chick embryo cells vidarabine also inhibited the multiplication of Rous sarcoma virus, an RNA virus which depends upon cell DNA synthesis at a stage in its replicative cycle. It will be shown below that vidarabine is metabolized to its hypoxanthine analogue by deamination, and this metabolite was found to possess antiviral activity, but only to the extent of about 10% of that of vidarabine.

Vidarabine was also found active against herpes virus infections in animals. There was some increase in the number of survivors and prolongation of survival time in mice infected intracerebrally and treated with vidarabine by oral, intraperitoneal or subcutaneous routes. The activity was not great, however, and daily doses of up to 2000 mg/kg were required to achieve a 91% increase in the number of surviving animals. Inoculation of herpes virus by corneal scarification in hamsters caused keratitis, leading by central spread of virus to a fatal encephalitis. The number of survivors could be significantly increased by applying 5–20% vidarabine ointment to the eyes twice daily. Some effect in herpetic encephalitis was also observed in mice infected with the virus intracerebrally, but large doses of up to 2000 mg/kg daily were necessary to achieve a 90% increase in the number of survivors. Vidarabine was also active in mice infected with vaccinia virus by the intracerebral route, but here again high doses were required. Daily doses of 1000 mg/kg given by mouth gave six of ten survivors, compared with none of ten untreated control animals, and there was also some increase in survival time. A similar result was obtained with doses of 250–1000 mg/kg given intraperitoneally.

Vidarabine had some inhibitory effect against myxoma and pseudorabies viruses (Sidwell *et al.*, 1970), and herpesvirus simiae, a virus of the herpes group which causes malignant lymphoma and lymphocytic leukaemia in owl monkeys (Adamson *et al.*, 1972).

The antiviral spectrum of vidarabine is shown in Table 3.6.

Table 3.6 Reported antiviral spectrum of vidarabine

Type	Group	Virus
DNA	Herpesvirus	Herpes 1
		Herpes 2
		Pseudorabies
		Herpesvirus saimiri
		Varicella-zoster
		Cytomegalovirus
	Poxvirus	Vaccinia
		Myxoma
RNA	Leukovirus	Rous sarcoma

Toxicity

Some information on the toxicity of vidarabine was obtained in the initial studies of its antiviral activity (Miller *et al.*, 1969). Cultures of HEp-2 cells exposed to vidarabine for 72 hours showed evidence of toxicity at 160 µg/ml, but not at 80 µg/ml. The cell count was reduced at the higher concentration, but growth was not completely arrested. The antiviral action of vidarabine was evident with concentrations of 40 µg/ml or lower, and was therefore not an non-specific effect due to toxicity.

The toxicity was studied in some detail by Kurtz *et al.* (1968). The intraperitoneal LD50 in mice was 4677 mg/kg. The animals showed depression, and severe abdominal irritation. The peak of mortality occurred on the first day, and the survivors showed various degrees of debility and weight loss over an observation period of 28 days. No signs of toxicity were seen in mice given a single dose of 7950 mg/kg orally. When vidarabine was incorporated in the diet to the extent of 2% the mice showed a mean loss of weight of 30% after 2 weeks, and some died. The intake of food was reduced. Similar effects were noted with concentrations in the diet down to 0·25%, and only at a level of 0·1% in the diet did the animals show a normal weight gain. Examination of the blood showed an increase in neutrophils with lymphopenia at all dietary levels, and there was also an increase in the haematocrit and haemoglobin levels. At autopsy the spleen was markedly reduced in weight, and there was a variable reduction in the weight of the liver. At the highest dose level the hepatocytes were enlarged and vacuolated, and contained large accumulations of glycogen which compressed the formed elements of the cytoplasm. There was also atrophy of the thymus.

Application of 3·3% and 10% vidarabine ointments to the skin of rabbits caused marked erythema, and transient erythema was noted after application to the eye. Pavan-Langston and Dohlman (1972) observed that patients being treated with 3·3% vidarabine ointment for herpetic keratitis could tolerate it for 2 months continuously. Three patients out of 23 developed mild drug reactions consisting of irritation and a burning sensation, but were nevertheless able to continue the treatment for 2 weeks. In one patient with an allergic response to idoxuridine ointment the symptoms disappeared when the treatment was changed to vidarabine ointment.

Toxic side-effects have been observed in patients with malignant conditions during treatment with intravenous infusions of vidarabine. In six patients given 10–15 mg/kg daily for 5–15 days the only side-effect observed was nausea, which was not severe enough to require cessation of treatment (Ch'ien et al., 1973 a). When higher doses were given toxic side-effects were much more marked. Bodey et al. (1974) treated 28 patients with daily infusions of 1 g/m^2 for 3–14 days, and observed nausea, vomiting and diarrhoea in 61%, leukopenia in 67% and thrombocytopenia in 89%.

Vidarabine will cause breakage of chromosomes. Nichols (1964) incubated cultures of human leukocytes in medium containing vidarabine and made chromosome preparations after 36 and 60 hours. With a concentration of 1×10^{-4} M breaks were seen in 32–46% of cells, and there were usually one or two breaks per cell. With 3×10^{-4} M up to 66% of the cells were affected, with several breaks per cell. The effect was not reversed by the normal deoxyribosides, and was thus ascribed to irreversible inclusion of vidarabine into the DNA molecule. Wilkerson et al. (1973) observed chromosome abnormalities in preparations from the blood of patients undergoing treatment with vidarabine. Thirty minutes after setting up an intravenous drip breaks were seen in 13% of metaphase plates. The peak incidence of 25% breakage occurred 18–24 hours after beginning treatment and was followed by a gradual decline. Four weeks after the end of treatment breaks were still seen in 9–10% of the metaphase plates.

Rao and Natarajan (1965) observed chromosome abnormalities in preparations from secondary root tips of Vicia faba exposed to 2 mM vidarabine for 6 hours; there were Feulgen-negative achromatic regions on one or both chromatids, chromosome gaps and chromatid breaks.

Vidarabine has little or no effect upon pre- and post-natal development. Adlard et al. (1975) gave it to pregnant rats at the 14th day of gestation, corresponding to the onset of the major period of neurogenesis in the embryo, and also to the progeny on the 5th day after birth. Prenatal growth was unaffected by doses to the mother as high as 1 g/kg, and there was no retardation in post-natal brain growth. Vidarabine given post-natally in doses of 50 and 250 mg/kg did not affect growth of the body, whole brain or cerebellum over an observation period of 25 days, but a 10% reduction in the weight of the cerebellum was seen in animals given four doses of 1 g/kg. Cytarabine given under the same conditions severely restricts pre- and post-natal growth of the brain. The evident resistance of the developing brain to vidarabine suggests that it would be preferable to cytarabine in the treatment of congenital cytomegalovirus infections. Similar observations were made by Fishaut et al. (1976). Litters of 2-day-old rats were injected intraperitoneally with vidarabine in doses of 3–50 mg/kg on the 2nd, 3rd, 4th and 5th days of life. Over an observation period of 60 days they did not differ from untreated litters in respect of survival, weight gain, vigour and developmental motor behaviour.

Metabolism

The evidence for and against phosphorylation of vidarabine and subsequent incorporation has been discussed in the section on the mode of action. The metabolic degradation of vidarabine in vitro was studied by Drach et al. (1973). Tritiated vidarabine labelled in the 2- position was incubated with a rat liver extract. The label was almost completely released as tritiated water, and the overall reaction was blocked by xanthine oxidase inhibitors, such as allopurinol. When mice were injected with labelled vidarabine intraperitoneally the urine contained only the hypoxanthine analogue, indicating the occurrence of deamination, and vidarabine was not present as such. Deamination was carried out by extracts of liver and other organs, but no cleavage to the free base could be detected (Brink and LePage, 1964 b). Borondy et al. (1973) observed species differences in the metabolic degrada- tion of vidarabine. When labelled with tritium in the 2- position, the urinary excretion of non-volatile tritium, indicating intactness of the ring system, was 10–20% of the dose in rats, 20% in dogs, 25–45% in mice and 90–95% in rhesus monkeys. In patients 40–70% was excreted over a 5–10 day collection period. The major metabolic product in plasma, urine and animal tissues was the hypoxanthine analogue, and considerable amounts of tritiated water were found. When vidarabine labelled with ^{14}C in the 8- position was given to rats by mouth 18% of the radioactivity was recovered in the breath as CO_2, a finding which indicated that some breakdown of the imidazole ring had occurred.

Clinical pharmacology

Some account of the behaviour of vidarabine in animals and man has already been given in the preceding section. Brink and LePage (1964) also investigated the distribution and excretion of labelled vidarabine in mice. After intraperitoneal administration it was rapidly absorbed into the blood, with a peak level per ml of 3% of the dose after 30 minutes. After intravenous administration it was first concentrated in the red cells, but had equilibrated with the plasma by 15 minutes; 84% of the dose was cleared from the blood in 1 minute and 95% by 30 minutes. In 60 minutes 35% of the radioactivity had appeared in the urine.

Similar studies were carried out in man by Ch'ien et al. (1973 b). After giving tritiated vidarabine intravenously in a dose of 1 mg/kg the total plasma tritium level was equivalent to 1·4 µg/ml of vidarabine after 30 minutes, but by 8 hours this had dropped to an undetectable level with a half-life of 1·5 hours. It was excreted in the urine to the extent of 16% of the dose in the first 4 hours, mainly as hypoxanthine. The rate of excretion then decreased, and 0·1% of the dose was excreted between 16 and 24 hours. At the same time as this decrease there was a proportionate increase in the excretion of degradation products, indicated by an increase in the proportion of volatile tritium.

The metabolism of vidarabine in man was slower than in lower mammals, and resembled that in the monkey. It was inferred from these observations that intravenous infusion is the preferred method of administration, and treatment for 12 hours daily should not result in excessive accumulation in view of the rapidity of excretion. Further observations were made by LePage et al. (1973) in patients with advanced carcinoma. Two patients who received a dose of 500 mg/m^2 intramuscularly excreted only 1·7% and 7·0% of the dose, a finding which indicates that vidarabine is poorly absorbed when given by this route. Two patients who were given rapid intravenous injections of 250 mg/m^2 excreted 88% and 97% of the dose in 24 hours. A peak plasma level of around 5 µg/ml was attained in one patient after 6 hours, and the other had continuous low levels of around 1 µg/ml; both had high initial levels of the hypoxanthine analogue of up to 22 µg/ml. Two patients given a continuous drip of 544 mg/m^2 and 382 mg/m^2 had continuous and rising levels, one of around 3–4 µg/ml of both vidarabine and its hypoxanthine analogue, and the other of around 5–8 µg/ml of the analogue with levels of vidarabine below 1 µg/ml. Vidarabine and the hypoxanthine analogue accounted for more than 90% of the radioactivity found in the plasma and urine up to 6 hours after administration, and 65–85% of radioactivity in urine collected between 6 and 24 hours. Glasko (1973) determined tissue levels of vidarabine in autopsy specimens from a patient who died 11 days after a single intravenous dose containing an unspecified amount of tritiated vidarabine. From calculations based on radioactivity the plasma contained 0·24 µg/ml, the kidney 0·56 µg/ml, spleen 0·53 µg/ml, liver 0·51 µg/ml, heart 0·32 µg/ml, brain 0·17 µg/ml and skeletal muscle 0·14 µg/ml. The level in the brain was thus not much lower than that in the plasma. In another patient examined 2 hours after dosing the plasma contained 0·48 µg/ml and the cerebrospinal fluid 0·24 µg/ml. The activities recorded would be largely accounted for by metabolites, but these findings are important in showing that vidarabine can cross the blood–brain barrier and is therefore suitable for consideration in the treatment of herpetic encephalitis.

As vidarabine is used for the treatment of herpetic keratitis its clinical pharmacology in the eye is of great importance. Pavan-Langston *et al.* (1973) carried out penetration studies in 80 rabbits treated with 3% vidarabine eye ointment in various bases, or given 25 mg or 100 mg vidarabine subconjunctivally. Samples of aqueous were taken at intervals up to 24 hours. Vidarabine was not found after any method of administration, but significant amounts of the hypoxanthine analogue were found. The highest concentration was 20 µg/ml, achieved with 3% vidarabine in water-miscible cream base. Examination of the corneas of the rabbits by slit-lamp and electron microscopy showed no evidence of toxicity. There was good preservation of the surface microplicae and epithelium as a whole, with no disturbance of basement or Bowman's membrane and stroma. Vidarabine ointment was used on 16 patients with chronic herpetic keratouveitis resulting from treatment with idoxuridine and steroids. It was found in the aqueous, together with the hypoxanthine analogue, in only two patients, both of whom had large epithelial defects which presumably allowed penetration to occur. The maximum amounts found were 1·0 µg/ml vidarabine and 1·0 µg/ml of the analogue.

Assay methods

It should be possible to determine the concentration of vidarabine by biological methods of the kind used with idoxuridine and cytarabine, but no reports have so far appeared. In the studies of Pavan-Langston *et al.* decribed in the preceding section, vidarabine was determined by analytical chromatography in a Varian Aerograph LC4100 liquid chromatograph with two high pressure pumps with an inlet pressure of 1 ton/in^2, a gradient generator, a u.v. detector working at 254 nm and a constant temperature bath. Working details of the method were not given.

Clinical use

Vidarabine is indicated for the treatment of infections caused by viruses of the herpes group. It has also been used in the treatment of smallpox.

Contraindications

No information is available on contraindications, but it would be advisable to restrict the systemic use of vidarabine during pregnancy to the treatment of infections associated with a high mortality.

Table 3.7 Properties of antiviral nucleosides

Compound	Base	Pentose	M.W.	Solubility (%)	Half-life (minutes)	Crosses blood–brain barrier	Teratogenic	Immuno-suppressive
Cytarabine	Pyrimidine	Arabinose	243·2	10	30–60	Not reported	Yes	Yes
Idoxuridine	Pyrimidine	Ribose	354·1	0·8	—	No	Yes	Yes
Trifluorothymidine	Pyrimidine	Deoxyribose	296·2	⩾ 5	30	Not reported	Yes	Not reported
Vidarabine	Purine	Arabinose	267·3	0·05	90	Yes	No	Not reported

Preparations

No preparations of vidarabine are available at present. The pure substance is obtainable from Parke, Davis and Co.

The main properties of the antiviral nucleosides are summarized in Table 3.7.

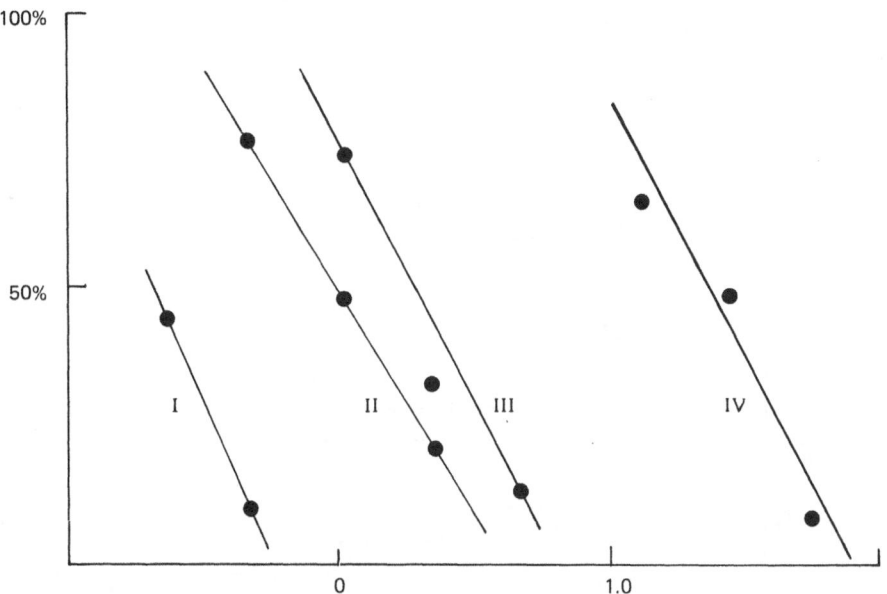

Figure 3.4 Dose–response lines of the antiviral nucleosides against a type 1 strain of herpes virus. Ordinate: mean plaque number as a percentage of the control value. Abscissa: concentration of compound ($\log_{10}$ µM). I, Cytarabine; II, Idoxuridine; III, Trifluoro-thymidine; IV, Vidarabine

RELATIVE POTENCIES

Their relative potencies cannot be readily ascertained from the literature, since the compounds have not been compared in uniform test systems. This point was investigated by Collins and Bauer (1977), who obtained dose-response lines in VERO cells by the plaque reduction method. The relative potencies against a strain of type 1 herpes virus are shown in Figure 3.4. Cytarabine is by far the most active compound, with an ED50 of 0·25 µM.

Idoxuridine and trifluorothymidine are somewhat less active (1 µM and 1·6 µM), whereas the activity of vidarabine is relatively low (18 µM). The clinical effectiveness of the latter compound in herpetic keratitis is evidently due to the fact that it is applied topically; in spite of the low solubility a concentration of around 2000 µM can be attained locally.

The relative potencies of the nucleosides against the two types of herpes virus are shown in Figure 3·5, which shows the ED50 values of dose–response lines against five strains of each type. There is some overlap in sensitivity, but there is a general tendency for type 2 strains to be less sensitive in each case.

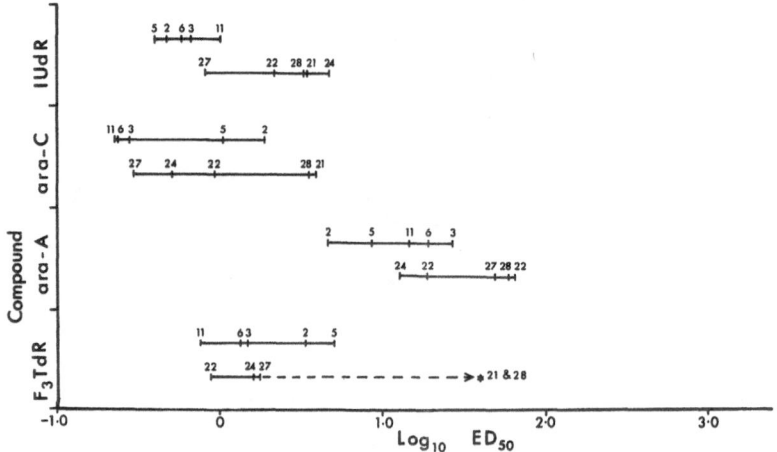

Figure 3.5 Relative sensitivities of 5 type 1 and 5 type 2 strains of herpes virus to the four antiviral nucleosides. Abscissa: $\log_{10}$ ED50 (µM). IUdR: idoxuridine. Ara-C: cytarabine. Ara-A: vidarabine. F_3TdR: trifluorothymidine. The numerals are reference numbers of the strains. The bars represent the ED50 values. In each pair of lines the upper represents type 1 strains and the lower line type 2 strains. * Strains 21 and 28 were not inhibited by trifluorothymidine in subtoxic concentrations

The relative potencies against vaccinia virus are shown in Figure 3.6. Cytarabine is again the most active compound, with an ED50 of 4 µM, followed by vidarabine (7 µM). Trifluorothymidine (125 µM) and idoxuridine (160 µM) are much less active, but their solubility is high enough to permit effective concentrations to be obtained in eye drops in the treatment of vaccinial keratitis.

The antiviral nucleosides have found a valuable place in the treatment of superficial infections with herpes and vaccinia viruses, but their systemic use, as in the treatment of herpetic encephalitis, generalized herpes and smallpox, is limited by their short half-lives and relatively high toxicity.

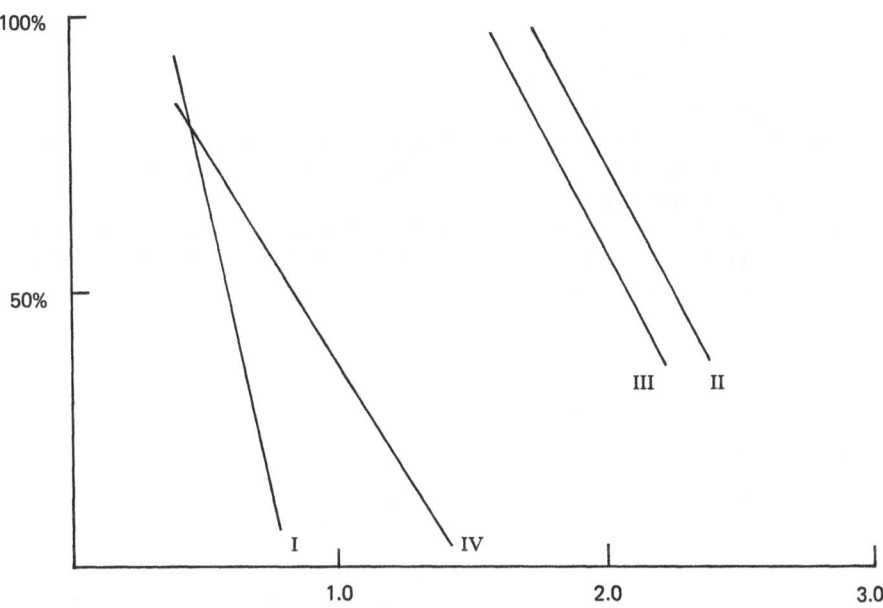

Figure 3.6   Dose–response lines of the antiviral nucleosides against the Lister strain of vaccinia virus. Ordinate: mean plaque number as a percentage of the control value. Abscissa: concentration of compound ($\log_{10}$ μM). I, Cytarabine; II, Idoxuridine; III, Trifluorothymidine; IV, Vidarabine

CHANGE OF NOMENCLATURE

The literature on the antiviral nucleosides can be found in the indexes of *Chemical Abstracts* under the compound names used in this chapter as far as 1972. From then onwards a different nomenclature is used in some cases, and in searching for references for this and subsequent years the compounds should be located in the indexes under the following names:

Cytarabine	2(1*H*)-Pyrimidinone, 4-amino-1-β-D-arabinofuranosyl-
Idoxuridine	Uridine, 2′-deoxy-5-iodo- (name unchanged)
Trifluorothymidine	2,4(1*H*,3*H*)-Pyrimidinedione, 1-β-D-ribofuranosyl, 3-trifluoromethyl-
Vidarabine	9*H*-Purine-6-amine, 9-β-D-arabinofuranosyl-

GENERAL READING

Davidson, J. N. (1972). *The Biochemistry of the Nucleic Acids*. London: Chapman and Hall

Prusoff, W. H. and Goz, B. (1973). Chemotherapy—molecular aspects. In: A. S. Kaplan (ed.) *The Herpesviruses* 641–663, (New York and London: Academic Press Inc.)

Schabel, F. M. and Montgomery, J. A. (1972). Purines and pyrimidines. In: D. J. Bauer (ed.) *Chemotherapy of Virus Diseases, Vol. 1*, 231–363, (Oxford: Pergamon Press)

Chapter 4

Antiviral agents—II

Amantadine hydrochloride and methisazone

AMANTADINE HYDROCHLORIDE

Amantadine is the generic name of 1-adamantanamine. It is also referred to in the literature as 1-aminoadamantane. It is a primary amine of a saturated hydrocarbon which has a cage structure (Figure 4.1). It is used as the hydrochloride, which is a stable white crystalline powder soluble in water to the extent of 40%. The compound is distinctly volatile (Vernier *et al.*, 1969).

In addition to amantadine its derivative Rimantadine (α-methyl-1-adamantanemethylamine hydrochloride) and also cyclooctylamine have been successful in clinical trials. They have as yet not been brought into general use, however, and will not be discussed further in this section.

Biological actions

In early work on the effects of amantadine in the whole animal it was found that intravenous doses of 5–20 mg/kg caused slowing of the heart rate and extra systoles in rabbits. Doses of 30–50 mg/kg caused severe dysrhythmias, such as ventricular flutter, ventricular fibrillation and irreversible conduction block. Prolongation of the P–Q interval and duration of the QRS complex occurred with doses of 10 mg/kg and higher (Kaji *et al.*, 1966). Wesemann and Zilliken (1967) found that amantadine hydrochloride in doses higher than 10 µM sensitized the rat fundus strip preparation to the motor effect of 5-hydroxytryptamine by a factor of 10–100. With higher concentrations the ability of the strip to contract in response to 5-hydroxytryptamine was lost and could not be restored by washing. The authors suggested that amantadine interacts with 5-hydroxytryptamine receptors in some way.

The chance observation that amantadine hydrochloride is effective in the treatment of parkinsonism (Schwab *et al.*, 1969) has led to an examination of its pharmacological properties in greater detail.

An increase in synthesis and release of dopamine was observed in slices of corpus striatum from rats injected subcutaneously with amantadine hydrochloride in a dose of 40 mg/kg 2 hours previously (Scatton *et al.*, 1970). A relationship with dopamine was also observed by Brelak *et al.* (1970). In dogs given a priming dose of 0·1 mg/kg dopamine intravenously the administration of amantadine hydrochloride 6–8 minutes later caused a dose-related rise in blood pressure, observable with a dose as low as 0·08 mg/kg. Amantadine hydrochloride alone had no pressor effect, except in very high doses. These results indicated that amantadine was releasing dopamine or other catecholamines from neuronal storage sites, and the authors suggested that this action might account for its effect in the treatment of parkinsonism. Strömberg *et al.* (1970) came to the same conclusion, and suggested further that amantadine hydrochloride might act predominantly on the central neurones. It also inhibited the rigidity and hyperactivity produced by oxotremorine in rats, a finding which is consistent with the view that it abolishes rigidity in parkinsonism by activation of cerebral dopaminergic function (Jurna *et al.*, 1972). Amantadine hydrochloride was found to potentiate the increase in motor activity produced in mice by L-dopa, a result which provides an explanation for the clinical observation that the administration of amantadine hydrochloride in parkinsonism allows the dose of L-dopa to be reduced (Svensson and Strömberg, 1970).

Other actions have been reported for amantadine hydrochloride which are unrelated to its interaction with catecholamines. Rawls *et al.* (1967) found that concentrations of 25 µg/ml or higher inhibited the transformation of human lymphocytes by *Phaseolus vulgaris* phytohaemagglutinin but did not affect their ability to be agglutinated. This observation suggests that amantadine hydrochloride interacts with the cell surface, and is in accordance with the finding, described in the section on mode of action, that the adsorption of influenza virus to the cell surface takes place in the presence of amantadine hydrochloride, but penetration into the cell interior is delayed. Bredt and Mardiney (1969) were unable to confirm this inhibitory effect on transformation by phytohaemagglutinin, but observed complete inhibition of transformation when allogeneic lymphocytes were used as the transforming agent.

Amantadine hydrochloride has a general inhibitory effect on cell growth. Balode and Gibadulin (1970) found that a concentration of 25 µg/ml reduced DNA synthesis in chick embryo fibroblasts by 24%. The generation time of the cells was lengthened from 32 to 108 hours, and the duration of DNA synthesis from 5·3–9·5 hours to 12·4–13·5 hours.

Antiviral activity and mode of action

The mode of action of amantadine hydrochloride is intimately connected with its activity against influenza virus, and the two topics will therefore be reviewed together. The mechanism of its pharmacological activities will not be considered further, although they need to be borne in mind when using amantadine hydrochloride as an antiviral agent.

The antiviral activity of amantadine hydrochloride was first observed by Davies *et al.* (1964), who found that it was active against certain members of the ortho- and paramyxovirus groups. In tissue culture it inhibited the multiplication of four strains of influenza A, 1 of A1, 3 of A2, 1 of influenza C, and also Sendai virus (a paramyxovirus). Other strains were unaffected, including two strains of influenza B, Newcastle disease virus, mumps, and parainfluenza 2 and 3. The compound was inactive against a general range of RNA and DNA viruses. In chick embryo fibroblasts strains sensitive to amantadine hydrochloride were maximally inhibited at a concentration of 25 µg/ml. In single-cycle experiments the maximum reduction in yield attained was 1 log unit (90%). No inactivation occurred when virus suspensions were exposed to concentrations of 25 µg/ml for 24 hours at 37 °C, but there was some inactivation at a concentration of 100 µg/ml. Amantadine hydrochloride did not prevent the uptake of virus by cells, or its release at the end of the replication cycle.

The concentration of amantadine hydrochloride used in the test in which it was found inactive against parainfluenza 2 and 3 was not specified. It may have been too low, since Cochran *et al.* (1965) found activity against these viruses in HeLa and LLC-MK2 cells when a concentration of 62·5 µg/ml was used.

Antiviral activity could also be demonstrated in fertile eggs and mice. The multiplication of A/PR8 (HON1) in the allantoic cavity was inhibited when 500 µg of amantadine hydrochloride was injected into the yolk sac. In mice infected intranasally with A/swine, A/WS and A2 strains some reduction in mortality and prolongation of survival time was obtained when amantadine hydrochloride was given in doses of 40 mg/kg by intraperitoneal, subcutaneous or oral routes, starting 30 minutes before infection and continuing every 4 hours for 48 hours.

The antiviral activity of amantadine hydrochloride was confirmed by Schild and Sutton (1965), who also noted a considerable variation in activity against different strains. There was little effect against the NWS, A/PR8 and A1 FM1 strains, whereas A2/Scot/49/57 and 442/63, a more recent A2 strain, were completely inhibited. Variation in sensitivity was also seen in five strains of animal origin, the most sensitive being A/equine/Miami/63.

Influenza virus undergoes the phenomena of antigenic shift and antigenic drift, but the strains now prevalent are still sensitive to the action of amantadine. In 1972 the new variant strain A/England /42/72 emerged from the prevailing A/Hong Kong/68 and caused widespread influenza in the winter of 1972–73. Tests in monkey kidney cells by McLaren and Potter (1973) showed that the new variant was as sensitive to amantadine hydrochloride as the Hong Kong strain, the mean inhibitory concentration ranging from 0·05 to 0·15 µg/ml (mean 0·10 µg/ml), compared with 0·03–0·10 µg/ml (mean 0·06 µg/ml) for the latter.

The activity of amantadine hydrochloride in mice described in early reports has been studied further. Some protective effect was seen in mice infected with the A2/Sing/4/57 strain intranasally and given amantadine hydrochloride intraperitoneally in a dose of 70 µg/kg daily for 5 days (Schild and Sutton, 1965). Grunert et al. (1965) infected mice intranasally with 3–6 LD50 of the A/swine/S15 strain of influenza virus and treated them with amantadine hydrochloride given intraperitoneally every 4 hours, beginning 4 hours before infection and continuing up to 43½ hours afterwards. In comparison with untreated control animals there was some increase in the percentage of survivors, which increased with increasing dose. The dose which increased the percentage survival to half-way between the control value and 100% (the corrected ED50) was 7 mg/kg. Amantadine hydrochloride was also effective when given by the oral route; a single dose given 1 hour before infection was effective, and the effect was not increased by continuing the treatment. No protective effect was observed if treatment was delayed until 24 hours after infection, or a single dose was given 9 hours before infection. In further work with the same strain of virus it was found that treatment with amantadine hydrochloride given intraperitoneally in a dose of 20 mg/kg every 4 hours, beginning 1 hour before infection, would reduce the virus content of the lungs by 1 log unit or so (Davies et al., 1966). McGahen and Hoffmann (1968) observed an apparent therapeutic effect in mice. In preliminary work they noted that the rate of water consumption of mice infected with the A2/Bethesda/10/63 and A2/AA/2/60 strains decreased 32 hours after infection, and this was considered to be an indication of the onset of clinical illness. When 0·5 mg/ml amantadine hydrochloride was added to the drinking water from the 48th to the 240th hour after an infection the proportion of survivors rose from 14·3% to 42·3% in a group of 78 animals. There was also an increase in the mean survival time and a reduction in the severity of illness. However, this effect was probably not antiviral in nature, since the virus would probably have attained maximum titre in the lungs by the time treatment was begun.

Resistance to amantadine hydrochloride can be readily produced by passage in the presence of the compound. Cochran et al. (1965) found that the A2/Japan/305 strain became completely resistant after one passage in calf kidney cells in the presence of an unstated concentration. Resistance can also be induced in mice (Oxford et al., 1970). The A2/Singapore/1/57 strain was passaged in mice which were given 50 mg/kg amantadine hydrochloride intraperitoneally 30 minutes before infection and maintained on drinking water containing 1 mg/ml. The sensitivity of the virus was determined at each passage by determining the concentration which would produce 50% inhibition of haemadsorption in BSC-1 cells. The initial virus had sensitivity values of 0·04–0·43 µg/ml (mean 0·3 µg/ml), and this had risen to 10–50 µg/ml (mean 25·3 µg/ml) after six passages. The resistance was not lost after four passages in eggs in the absence of amantadine hydrochloride.

The relation of dose to the induction of resistance was investigated by Oxford and Potter (1970). When passaged in mice treated with 1·5 and 15 mg/kg amantadine hydrochloride daily the A2/Singapore/1/57 strain remained sensitive, but resistant variants were readily obtained when the dose was increased to 150 mg/kg. In some cases a 500-fold increase in resistance was obtained. The resistant variants showed no change from normal in buoyant density, morphology and serology. The authors point out that the dose required to produce resistance was very much higher than the recommended human dose of 200 mg daily.

Goedemans and de Bock (1970) investigated some of the biological properties of a resistant strain of A/swine virus. Resistance was produced by serial passage of this strain in the allantoic cavity in the presence of 200 µg of amantadine hydrochloride. Adsorption on to chick embryo fibroblasts was retarded by a concentration of 40 µg/ml in the usual way in tests with the virus before passage and up to the third passage, but at the sixth passage adsorption was no longer affected. In a study of the penetration phase of the growth cycle the virus was adsorbed on to chick embryo fibroblasts, which were then exposed to a concentration of 40 µg/ml from 15 minutes to 3 hours after infection, and the amount of virus which had penetrated to the interior of the cell was determined by sonication and titration of haemagglutinin content. Entry into the cell was completely inhibited at the third passage, but resistance appeared at the sixth passage, and by the tenth passage the amount of virus gaining entry to the cell was only 1 log unit lower than in the case of the initial virus. Resistance was still retained after four passages in the absence of amantadine. Development of resistance also occurred *in vivo*. The A2/Japan strain was passaged in mice which were given 0·5 mM/kg amantadine base by mouth each day and tested for resistance in tissue culture at each passage level. Resistance to the action of amantadine hydrochloride during the penetration phase of the growth cycle appeared after six passages, but was still incomplete after ten. The resistance disappeared after three passages in eggs in the absence of the compound, and thus differed from the resistance which arose in egg passage. The authors calculated that 80 passages in the presence of the compound were required for the development of resistance. Resistance is thus unlikely to appear when amantadine hydrochloride is used clinically as an antiviral agent.

Amantadine hydrochloride is active against rubella, a virus which is unrelated to the myxoviruses and regarded by many as a member of the togavirus group. The activity was first reported by Maassab and Cochran (1964). Cultures of the LLC MK2 continuous line of monkey kidney cells were exposed to amantadine hydrochloride in a concentration of $31 \cdot 2$ µg/ml and infected with the WM strain of rubella virus 5 minutes later. After an adsorption period of 5 hours the residual virus inoculum was removed and replaced with medium containing the compound in the same concentration. The virus content of the supernatant was determined after 3 and 6 days by a method based upon the interference effect between rubella and ECHO 11 virus. There was a reduction in yield of 2 log units in comparison with un-treated cultures. Amantadine hydrochloride was effective when present during the first 3 hours after infection, but when the time of addition was delayed to 5 hours after infection no inhibition of virus growth was obtained.

The activity against rubella was studied further by Oxford and Schild (1965). In experiments with five strains of rubella virus growing in RK-13 cells amantadine hydrochloride produced some suppression of cytopathic effect in comparison with untreated cultures. No suppression occurred with concentrations lower than 10 µM, and for complete suppression a concentration of 40 µM was required. A study of the growth cycle of rubella virus showed that infective virus began to appear 18 hours after infection, and attained a maximum titre at 120 hours. In the presence of amantadine hydrochloride infective virus appeared after the same interval, but the maximum titre attained was about 2 log units lower, and no alteration in the relative proportions of extracellular and intracellular virus could be detected.

Attempts to demonstrate an antiviral effect against rubella in animals have been unsuccessful. Rhesus monkeys infected with rubella virus paren-terally or intranasally develop viraemia, shed virus from the pharynx and develop antibody, and this subclinical infection can be used as a model for testing antiviral activity *in vivo* (Stephenson *et al.*, 1965). Amantadine hydrochloride was given by mouth in doses of 75 mg/kg daily for 4–11 days to 13 monkeys, which were infected intravenously or intranasally with $2 \cdot 75$–4 log TCD50 of virus; 13 other monkeys were infected and left untreated as controls. No differences could be found in the three criteria of assessment between the treatment and control groups, and amantadine hydrochloride was thus ineffective against rubella *in vivo*. Oxford and Schild (1967 a) were unable to find any antiviral effect in hamsters and rabbits.

Amantadine hydrochloride has been found to inhibit a number of other unrelated viruses. There is a single unconfirmed report of activity against pseudorabies (Neumayer *et al.*, 1965). This is a virus of the herpes group and is the only DNA virus reported to be sensitive to amantadine hydro-chloride. Activity against the arenavirus group is better documented.

Monolayers of BHK-21 cells were infected with lymphocytic chorio-meningitis virus, a member of this group, and treated with amantadine hydrochloride from 2 hours before infection until 24 hours after. The yield of intra- and extracellular virus at the end of the growth cycle was reduced by concentrations of amantadine hydrochloride higher than 10 µg/ml, and at 50 µg/ml a 100-fold reduction was obtained (Welsh et al., 1971). Similar results were obtained by Pfau et al. (1972), who found that amantadine hydrochloride inhibited the multiplication of Amapari, Parana, Tacaribe and Tamiami viruses, which are also members of the arenavirus group.

Amantadine hydrochloride inhibits the multiplication of certain tumour viruses. Wallbank et al. (1966) found that the development of foci of tumour cells was inhibited in monolayers of chick embryo fibroblasts pretreated with amantadine hydrochloride, infected with Rous sarcoma virus and incubated in the presence of an overlay containing the compound. A similar result was obtained with Esh sarcoma virus. Neumayer et al. (1965) had previously found that amantadine hydrochloride was ineffective against Rous sarcoma virus, but in their test system the virus was inoculated on to the chorioallantoic membrane of fertile eggs and the compound was inoculated into the yolk sac, in which case the surface of the infected cells could not have been accessible to the compound. Essentially similar results were obtained by Oker-Blom and Andersen (1966). The spectrum of inhibition has also been extended to murine sarcoma virus (Rhim et al., 1972) and fowl leukosis virus (Oker-Blom and Andersen, 1967).

The basis of the mode of action of amantadine hydrochloride resides in the fact that it is a primary amine, and thus a derivative of ammonia. It had been known since 1961 that ammonium ions would inhibit the multiplication of influenza virus in tissue culture (Jensen et al., 1961). This resulted from the chance observation that the cytopathic effect produced by A/PR8 virus in dog kidney cells was inhibited in cultures which were alkaline. Further work showed that ammonium ion was the active agent, and inhibition could be obtained by the addition of ammonium chloride to the cultures. The minimum inhibitory concentration was 16 µg/ml, and at 40 µg/ml the yield of infective virus was reduced by 1·3 log units. Ammonium ions had no direct virucidal effect, since on incubation of the virus with 100 µg/ml ammonium chloride at 37 °C no inactivation occurred over a period of 5 hours. Ammonium chloride also did not interfere with the adsorption of the virus to the cells. In further work inhibition was observed in other tissue culture systems, and also with A1 and A2 strains. No protective effect could be obtained in mice infected with A/PR8 virus intranasally, presumably due to metabolism to urea preventing the access of ammonium ions to the infected cells.

Ammonium ions did not inhibit other myxoviruses (mumps, Newcastle disease virus and parainfluenza 3), or a range of other unrelated viruses

(Jensen and Liu, 1961). The antiviral activity of ammonium ions was reported independently by Eaton and Scala (1961), who found that concentrations of 5–13 µg/ml inhibited the growth of A/PR8 and Newcastle disease viruses in ascites tumour cells. They made the additional observation that there was little or no inhibitory effect when the addition of ammonium ions was delayed until 1–4 hours after infection. Jensen and Liu (1963) further showed that similar inhibition could be obtained with aliphatic amines, which may be regarded as substituted derivatives of ammonia. In cultures of dog kidney cells infected with A/PR8 inhibition of cytopathic effect and reduction in the yield of virus was obtained with primary aliphatic amines of chain lengths increasing from methylamine to butylamine. The minimum inhibitory concentration fell with increasing chain length, and a similar but less marked effect was obtained with secondary and tertiary amines. In a time of addition study carried out with propylamine the inhibitory effect was lost when addition was delayed until 1 hour or more after infection.

The discovery of the antiviral effect of amantadine hydrochloride by Davies et al. (1964) would appear to be a logical extension of the foregoing work on ammonia and amines. The mode of action was evidently the same, since the compound did not inactivate the virus on contact, did not affect adsorption, and only exerted a maximum inhibitory effect when added from 5 minutes to 1 hour after infection. In a study of the penetration of virus into chick embryo fibroblasts carried out with the A2/Japan/305 strain it was found that infected cultures treated with specific antibody $1\frac{3}{4}$ hours after infection produced essentially as much virus on further incubation as cultures not treated with antibody, a result which showed that the virus had already penetrated the cells by $1\frac{3}{4}$ hours and was therefore not accessible to neutralization by antibody. In the presence of amantadine hydrochloride, however, addition of antibody after $1\frac{3}{4}$ hours caused a reduction in virus yield of around 1 log unit. In the presence of the compound, therefore, much of the virus was still present at the cell surface and was neutralized by antibody. These observations were confirmed by Hoffmann et al. (1965) in experiments with the A2/AA/2/60 strain. Fletcher et al. (1965) observed a similar effect with ammonium ions and primary aliphatic amines. In cultures of monkey kidney cells infected with the A/PR8 strain the adsorbed virus became inaccessible to neutralization by 45 minutes, but in the presence of 250 µg/ml ammonium chloride the virus remained neutralizable for up to $3\frac{3}{4}$ hours. Similar results were obtained with propylamine, di- and triethylamine and other amines, and indicated that penetration of virus into the cell was being inhibited. The action of ammonium ions and amines thus resembled that of amantadine hydrochloride in an additional respect, and the authors concluded that all these substances had a common mode of action.

The mechanism of antiviral action against rubella appears to be the same as against influenza. Oxford and Schild (1965, 1967 b) showed that amantadine hydrochloride did not inactivate rubella virus *in vitro* and did not affect the rate of adsorption on to RK-13 cells; the multiplication of the virus was also inhibited by ammonium ions, and the virus was similarly not inactivated when incubated with 200 µg/ml ammonium acetate for 4 hours at 37 °C.

There is some evidence that the antiviral action of amantadine hydrochloride cannot be entirely explained by its reported effect in blocking or retarding penetration. The fever produced in rabbits by the intravenous injection of the A2/Singapore/1/57 strain of influenza virus was not prevented by an injection of 60 mg of amantadine hydrochloride given from 10 minutes to 24 hours previously. This result implies that the compound does not block the penetration of virus into lymphocytes, which then causes the liberation of the pyrogen (Grossgebauer and Langmaack, 1970). Kato and Eggers (1969) devised a method for studying the uncoating phase of the growth cycle in chick embryo cells infected with fowl plague virus. The virus was labelled with Neutral Red by carrying out five passages in cultures containing the dye. This procedure rendered the virus susceptible to photoinactivation. In cells infected with the labelled virus 90% of the inoculum became resistant to photoinactivation by 1 hour as a result of uncoating, but in the presence of 25 µg/ml amantadine hydrochloride this proportion was reduced to 30%. Amantadine hydrochloride thus had the additional action of inhibiting the uncoating process. Long and Olusanya (1972) studied some aspects of the molecular biology of the growth cycle of fowl plague virus. The synthesis of virus-directed RNA as measured by the uptake of tritiated uridine was inhibited by 25 µg/ml amantadine hydrochloride. It had previously been observed by Long and Burke (1969) that cycloheximide inhibited an early event in the growth cycle, probably the synthesis of RNA polymerase coded for in the RNA of the infecting particle. When infected cultures were exposed to cycloheximide, washed, and then exposed to amantadine hydrochloride, the multiplication of the virus was not inhibited. The amantadine-sensitive step had therefore taken place during the period of contact with cycloheximide. These findings agree with the work of Kato and Eggers, and localize the site of action of amantadine hydrochloride to the processes of uncoating and release of virus RNA.

Further information on the mode of action of amantadine hydrochloride has come from studies of its effect on arenaviruses. Welsh *et al.* (1971) found that the mechanism of action against lymphocytic choriomeningitis virus was essentially the same as for the myxoviruses and rubella, in that the compound did not inactivate the virus on contact, did not inhibit adsorption to cells, and delayed penetration, but they also observed that it appeared to act at two stages in the growth cycle, since the yield of virus was still reduced when the addition of amantadine hydrochloride was delayed for as long as 20 hours after infection. This could be explained by the observation that the compound becomes attached to the plasma membrane of the cell (Greenhalgh and Gaush, 1970), from which the virus particles emerge by budding (Dalton *et al.*, 1970). A two-stage mechanism of action was also observed by Pfau *et al.* (1972) in studies with other members of the arenavirus group.

Toxicity

The acute LD50 of amantadine hydrochloride for mice was reported by Davies *et al.* (1964) as 233 mg/kg intravenously and 1080 mg/kg by the oral route. Grunert *et al.* (1965) obtained values of 97 mg/kg and 700 mg/kg for the two routes respectively, and 205 mg/kg by intraperitoneal and 271 mg/kg by subcutaneous injection. When 12 doses were given with intervals of 4 hours the LD50 was similar by the various routes, and ranged from 147 mg/kg to 171 mg/kg. In rabbits the acute LD50 by intravenous injection was 30–40 mg/kg (Kaji *et al.*, 1966). Prokhorova and Solov'ev (1967) found that a single dose of 800 mg/kg given to mice by mouth caused vasomotor collapse and lesions in the liver parenchyma and mucosa of the small intestine. No pathological changes were seen in the internal organs after a single dose of 320 mg/kg or repeated doses of 128 mg/kg given by aerosol over a period of 10–14 days.

The chronic toxicity of amantadine hydrochloride was studied by Vernier *et al.* (1969). It was well tolerated when given to dogs by mouth over a period of 6–24 months in doses 13–33 times greater than those proposed for clinical use, and there were no pathological changes in the organs. Doses of 93 mg/kg caused stimulation of the central nervous system, with tremor, myoclonus, chronic convulsions, salivation, loss of light reflex, mydriasis and vomiting. Schwab *et al.* (1969) observed side effects in 36 (22%) of 163 patients with parkinsonism treated with amantadine hydrochloride, mainly insomnia, abdominal discomfort, anorexia, dizziness and depression. The effects disappeared within 36 hours of ending treatment. Amantadine hydrochloride also potentiated the effect of atropine-like drugs given in the treatment of parkinsonism, causing confusion and hallucinations.

The toxic effects of amantadine hydrochloride have mostly been associated with the use of doses considerably higher than that recommended for clinical use, which is 200 mg daily. At this dose level Jackson *et al.* (1967) found that side effects occurred in less than 20% of healthy volunteers. When the dose was increased to 400 mg daily side effects occurred in 40%, and with still higher doses nearly all subjects experienced adverse reactions. The usual effects reported were alterations in emotional state and cerebration, such as depression and inability to concentrate, and occasionally feelings of depersonalization and alterations in body image accompanied by anxiety. The symptoms appeared 3 hours after taking the drug, and usually passed off after 6 hours if the dose was not repeated. In a trial of amantadine hydrochloride carried out among elderly patients in a hospital for chronic diseases there was an increase in mortality in comparison with other patients. There were 26 deaths among 147 patients who took 200 mg a day for $11\frac{1}{2}$ weeks, giving a weekly mortality of 1·5%. The incidence among 600 other patients during the 4 weeks preceding the trial was 1·1%. During the trial period there were 11 deaths (0·3%) among 289 given placebo and eight (1·2%) among 60 left untreated. The mean control incidence was 0·5%, and differed significantly from the incidence among treated patients. The degree of comparability of the various groups was uncertain, but the authors nevertheless consider that the use of amantadine hydrochloride may carry a higher risk among elderly patients.

Tyrrell et al. (1965) reported the occurrence of troublesome insomnia among an unstated number of volunteers who took a daily dose of 400 mg, which made it necessary to reduce the dose to 200 mg. Information on the incidence of side effects is given in a number of reports of prophylactic and therapeutic trials, and is summarized in Table 4.1. In many cases the authors merely state that they observed no untoward reactions attributable to the drug. In several trials the incidence of side effects was specifically recorded, and it is evident that they appeared with similar frequencies in both treatment and placebo groups. The most frequently recorded symptoms were insomnia, headache, dyspepsia, nausea, vomiting and diarrhoea. One would expect to find symptoms of this nature in any large group of persons observed over a fair period of time. In therapeutic trials, in which patients were treated with amantadine hydrochloride when they were already suffering from influenza, it is difficult to distinguish between side effects and the usual symptoms of the illness. It may be concluded that amantadine hydrochloride can be given over extended periods without troublesome side effects if the recommended daily dose of 200 mg is not exceeded.

Metabolism

The metabolism of amantadine hydrochloride varies considerably in different species (Bleidner et al., 1965). When single oral doses ranging from 1 mg/kg to 100 mg/kg were given to mice, 51–75% of the dose was recovered in the urine unchanged. Some of the remainder was probably excreted as metabolites, since additional peaks could be found by gas chromatography, and some degree of metabolism was also implied by the incompleteness of recovery of the amount administered. Only small amounts were present in the faeces. In rats the recovery in the urine was only 16–18%. The recovery was also low in dogs, and N-methylamantadine was found as a metabolite, in amounts not exceeding 10% of the amount

Table 4.1 Reported side effects of amantadine hydrochloride

Reference	Dose (mg/day)	No. of subjects		No. with side effects		Side effects
		Treated	Placebo	Treated	Placebo	
Wendel et al. (1966)	200	469	380	8 (1·7%)	7 (1·8%)	Insomnia
Togo et al. (1968)	200	29	29	0	0	—
Wingfield et al. (1969)	100	23	48	0	0	—
Galbraith et al. (1969 a)	200	94	82	2	0	Headache
Galbraith et al. (1969 b)	200	102	100	3	0	Insomnia
Hornick et al. (1969)*	200	94	103	0	0	—
Knight et al. (1970)*	200	13	16	0	0	—
Togo et al. (1970)	200	54	48	0	0	—
Smorodintsev et al. (1970 a)	100, 200	206	198			Insomnia Dyspepsia
Smorodintsev et al. (1970 b)	100	1313	512	94 (7·1%)	26 (5·2%)	Headache
Oker-Blom et al. (1970)	200	192	199	—(8·7%)	—(3·4%)	Vertigo Insomnia Malaise
Kitamoto et al. (1970)*	200	182	173	18 (9·9%) 9 (4·9%) 5 (2·7%)	21 (12·1%) 9 (5·2%) 6 (8·5%)	Nausea Vomiting Diarrhoea
O'Donoghue et al. (1973)	200	50	61	0	0	Insomnia

* Therapeutic trial. Some reported side effects may have been symptoms of influenza

excreted as unchanged compound. In African green monkeys the recovery was 54%. In human volunteers 86% of a single oral dose was recovered unchanged in the urine; acetylated and methylated derivatives could not be detected, and there were no peaks indicative of metabolites in the gas chromatogram. Amantadine hydrochloride is thus readily absorbed from the intestines and is excreted without undergoing metabolic change.

Clinical pharmacology

The distribution of amantadine hydrochloride in mice was investigated by Uchiyama and Shibuya (1969), in experiments using compound labelled with tritium. It was given orally in a single dose of 1·6 mg/kg, corresponding to the dose in man recommended for clinical use. The compound attained maximum concentration in the internal organs 30 minutes after administration; the heart contained 0·3%, lungs 2·0%, liver 12·4%, kidney 5·2% and the spleen 0·7% of the dose given. Amantadine hydrochloride could be detected in the faeces and urine by 1 hour; by 12 hours 1·8% and 61·8% of the dose administered was excreted by the two routes respectively, and excretion continued up to 100 hours, when the total amounts were 7·5% and 86·7% respectively.

The blood level was 176 ng/ml at 15 minutes, 18·5 ng/ml at 30 minutes, 87 ng/ml at 1 hour and 45 ng/ml at 2 hours. Some compound was still present at 48 hours (14 ng/ml), but it could no longer be detected at 100 hours.

In three volunteers given single oral doses of 2–7 mg/kg Bleidner et al. (1965) found that the half-lives based on excretion were 9, 13 and 15 hours. After a dose of 5 mg/kg the maximum blood level was 0·6 µg/ml, and was attained 1–4 hours after administration. Biantrate et al. (1972) found blood levels ranging from 0·25 µg/ml to 1·01 µg/ml in four patients who were given 200–300 mg daily.

In volunteer studies carried out by Geuens and Stephens (1967) it was found that the rate of excretion of amantadine hydrochloride was highly dependent on the pH of the urine. The rate was very low when the urine was made alkaline by the administration of sodium bicarbonate, and increased rapidly when it was made acid by giving ammonium chloride. The body level of amantadine hydrochloride in a treated patient could thus vary widely, and a dosage sufficient for one patient might be insufficient for another. Toxic effects could be expected more frequently in patients who excreted the drug slowly. The authors suggested that combined treatment with sodium bicarbonate and amantadine hydrochloride would allow an effective level to be attained in 1 day, which could then be maintained on a reduced daily dose. They estimated that a loading dose of three times the daily maintenance dose would give a stable body level which would be three times as high if 4·2 g of sodium bicarbonate were given three times a day.

Combined treatment was not recommended for prophylaxis, but the authors saw no contraindication for using it in a 7-day course of treatment during the acute phase of influenza.

Assay methods

Amantadine hydrochloride is determined in blood, urine and tissue samples by utilizing its properties as an organic base, which enable it to be extracted into toluene and determined by gas chromatography. The method of Bleidner *et al.* (1965) is usually used, and amounts down to 0·1 μg can be determined. For the determination in blood samples 5 ml of oxalated whole blood is mixed with 20 ml of 5 N sodium hydroxide solution and 12·5 ml of benzene and shaken for 5 minutes. This procedure converts the amantadine hydrochloride to the free base which then passes into the benzene. One ml of 1 N hydrochloric acid is shaken with 10 ml of the benzene extract and the mixture is centrifuged to separate the aqueous phase, which now contains the amantadine as the hydrochloride. A volume of 0·75 ml of the aqueous phase, representing 3 ml of the original blood sample, is made alkaline with 1·5 ml of 1 N sodium hydroxide solution and the free base is extracted into 1 ml of benzene; 5 μl of this are injected into a gas chromatograph. The amount contained in it is determined by reference to standards prepared by adding known amounts of amantadine hydrochloride to blood and carrying out the same extraction procedure.

With urine samples the concentrations are higher and interfering substances are usually not present, so that the hydrochloric acid step may be omitted. The sample is mixed with 8 N sodium hydroxide solution and benzene; ratios of 4:4:2, 4:4:1 or 10:5:1 are suitable. The benzene extract is then used directly for gas chromatography. If interfering substances are present the method is modified by conversion of amantadine to its acetamido derivative. This is done by mixing 0·5 ml of benzene extract with 0·5 ml of acetic anhydride. The mixture is left overnight in a stoppered vessel. This is then chilled, and excess acetic anhydride is destroyed by the cautious addition of 2 ml of 8 N sodium hydroxide solution. An aliquot of the benzene phase is then analysed. Standards are prepared from known amounts of amantadine hydrochloride carried through the same procedure.

Samples of tissue are homogenized in five parts of 5 N sodium hydroxide solution and extracted into a measured volume of benzene. An aliquot of the benzene phase is then analysed. A similar method can be used for faeces.

Other workers have converted amantadine into a derivative as a standard procedure. Biantrate *et al.* (1972) used the trichloroacetyl derivative. In their method 1 ml of plasma is mixed with 1 ml of N sodium hydroxide solution and extracted with 5 ml of toluene, which is then extracted back with 2 ml of N hydrochloric acid. The aqueous phase is rebasified with

0·5 ml of 6 N sodium hydroxide solution and extracted with 2 ml of toluene; 20 µl of 2% trichloroacetyl chloride in toluene is added to 1·5 ml of the toluene extract and the mixture is heated at 70 °C for 30 minutes. Determination is carried out by gas chromatography of a volume of 1–3 µl. The method is linear over the range of 25–1000 ng/ml plasma. In a similar method Tsubouci *et al.* (1970) used the 1-monochloroacetamido derivative, and gave the sensitivity as 10 ng. They also developed a method based on a colour reaction which does not require gas chromatography. Amantadine reacts with *p*-nitrobenzaldehyde to give the 1-*p*-nitrobenzylidene derivative. This can be determined spectrometrically by its absorption at 292 nm. In biological fluids the method can be used down to 1 µg/ml.

Clinical use

Amantadine hydrochloride is used in the prophylaxis and treatment of influenza infections caused by A2 strains, and has been reported to shorten the duration of pain in zoster. It is indicated for the prophylaxis of influenza in persons who are at particular risk, such as infants and the elderly and patients suffering from chronic debilitating diseases, such as cardiovascular, renal and metabolic disorders.

Contraindications

Amantadine hydrochloride is contraindicated in patients with a history of epilepsy, elderly persons with cerebral arteriosclerosis and patients treated with stimulants of the central nervous system. It may enhance the effects of benzhexol, benztropine and orphenadrine, and if it is used it may be necessary to reduce the doses of these drugs. In view of its slight inhibitory effect on DNA synthesis it should not be used in pregnancy, although there are no reports of teratogenic effects.

It is not known whether amantadine hydrochloride passes into the milk, but it would be advisable to avoid its use during lactation.

Preparations

Amantadine hydrochloride capsules (USNF) 100 mg
Amantadine hydrochloride syrup (USNF)   50 mg in 5 ml
Symmetrel (Geigy); capsules of 100 mg.

CHANGE OF NOMENCLATURE

The literature on amantadine hydrochloride is indexed in *Chemical Abstracts* under 1-aminoadamantane as far as 1972. From this time onwards the literature should be looked for under the entry tricyclo[3,3,1,1^{3,7}]-decan-1-amine.

GENERAL READING

Hoffmann, C. E. (1973). Amantadine HCl and related compounds. In:
W. A. Carter (ed.) *Selective Inhibitors of Viral Functions* (Cleveland,
Ohio: Chemical Rubber Co. Press)

METHISAZONE

Methisazone (Figure 4.1) is the generic name of 1-methyl-1H-indole-
2,3-dione 3-thiosemicarbazone. It is also referred to in the literature as
1-methylisatin 3-thiosemicarbazone and N-methylisatin β-thiosemicarba-
zone. It is the 1-methyl derivative of isatin 3-thiosemicarbazone (isatin
β-thiosemicarbazone, IBT), and much of the work on the antiviral activity
and mode of action has been carried out with the parent compound, and
will be referred to in the sections which follow.

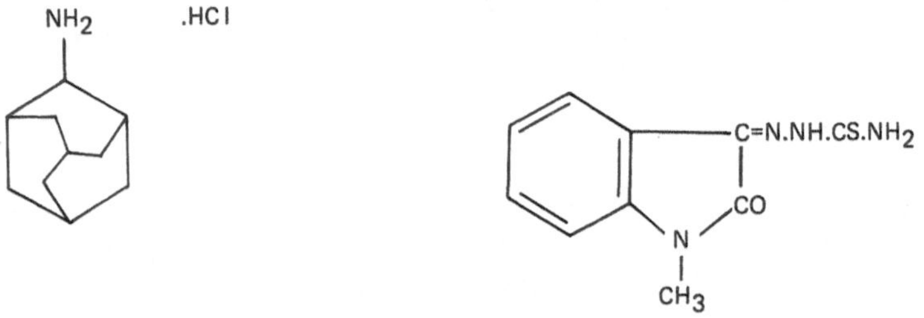

Figure 4.1   Structural formulae of (I) amantadine hydrochloride and (II)
methisazone

Methisazone is a yellow crystalline solid. It is soluble in water at room
temperature to the extent of 0·006%. As a result of this low solubility
methisazone can only be given by mouth. A stable 1% microcrystalline
suspension can be prepared by mixing 20 ml of a 5% solution of methisazone in dimethylformamide with 80 ml of water. A solution for use in
tissue culture can be prepared by dissolving 23 mg of methisazone in
0·5 ml of dimethylformamide and adding rapidly to 900 ml of distilled
water. The compound precipitates immediately in finely divided form, but
will go back into solution after autoclaving for 10 minutes at 10 lbs
pressure. Addition of 100 ml of a 10 × concentrated solution of tissue
culture salts gives 1000 ml of a 100 µM solution. This can be diluted
appropriately to give the concentration required. The solution should be

stored at 37 °C in the dark, but the compound will eventually crystallize out. Methisazone is soluble to the extent of at least 20% in dimethylsulphoxide, but the solution rapidly begins to deposit crystals and eventually solidifies. This is probably due to the formation of an adduct with the solvent, and the crystalline product is devoid of antiviral activity. Methisazone is therefore not suitable for local application in this vehicle.

Biological actions

The biological actions of methisazone, and perhaps also its antiviral action, are dependent to a considerable extent upon its chemical properties. The compound can exist in two isomeric forms, depending upon the configuration of the side-chain (Barz and Fritz, 1970). In the solid state it exists as the *syn*-isomer, with the oxygen atom in the 2- position forming a hydrogen bond with the hydrogen atom on the second nitrogen atom of the side-chain. In solution the compound assumes the *anti*-configuration, with the side-chain directed away from the remainder of the molecules. In this form the molecule acts as a bidentate ligand, and can form complexes with copper and zinc ions, which are co-ordinated with the two nitrogen atoms in the side-chain, and iron, cobalt and zinc co-ordinated with the first nitrogen and the sulphur atoms. The molecule can further assume a thiolate configuration, in which the hydrogen on the first nitrogen atom migrates to the sulphur atom. In this configuration the molecule can act as a tridentate ligand and form a divalent copper complex co-ordinated with the first nitrogen, oxygen and sulphur atoms.

In view of this chelating ability it is not surprising to find that methisazone is an inhibitor of certain metalloenzymes. Influenza A and influenza B contain an RNA-dependent RNA polymerase, which is associated with the core of the virion. The RNA polymerases of certain bacteria are known to be metalloenzymes containing zinc, and Oxford and Perrin (1974) were therefore led to study the effect of a number of chelating agents upon the influenza enzyme. In experiments with the RNA polymerase of the B/Lee strain a 50% reduction in activity was obtained with isatin 3-thiosemicarbazone in a concentration of 30 µM, and methisazone had the same inhibitory effect at 100 µM. When the sulphur atom is replaced by oxygen (isatin 3-semicarbazone) or an -NH group (isatin 3-amidinohydrazone) the chelating activity is much reduced, and the 50% inhibitory concentrations for these two compounds were above 500 µM. These observations led the authors to conclude that the RNA polymerase of influenza virus was probably a zinc metalloenzyme. Inhibition was also observed with the RNA polymerase of a strain of type A influenza (A/R1-5$^+$), but not with polymerases from *Escherichia coli* or *Micrococcus lysodeikticus*.

Haase and Levinson (1973) observed that methisazone inhibited the RNA-dependent DNA polymerase (reverse transcriptase) of the RNA-containing slow viruses visna, maedi and progressive pneumonia virus. A concentration of 40 µM caused an 85% reduction in enzyme activity. Levinson et al. (1973 b) observed a similar effect with the RNA-dependent DNA polymerase of Rous sarcoma virus. Methisazone in a concentration of 40 µM caused a 98% reduction in the activity of the enzyme. It also caused a 99% reduction in the transforming ability of the virus, as measured by counts of foci of transformed cells in chick embryo cell cultures. Concentrations of 1 µM reduced the respective activities to 42% and 50% of the control values, and to 1% in both cases when copper sulphate was also present in the same molar concentration. Cupric ions and methisazone were therefore acting in synergism. The authors suggest that the antiviral activity of methisazone and other thiosemicarbazones is a function of their ability to act as ligands for metallic ions. Steric and other factors must also be involved, however, since not all thiosemicarbazones possess antiviral activity.

In further work (Levinson et al., 1973 a) it was found that 40 µM methisazone would inhibit the RNA-dependent DNA polymerases of mouse sarcoma, mouse leukaemia and mouse mammary tumour viruses. Translation from single-stranded RNA was unaffected, however, since methisazone did not inhibit polypeptide synthesis from the messenger RNA of rat hepatoma cells.

Webb et al. (1965) found that the cyclic AMP content of human peripheral blood lymphocytes was increased 4-fold by incubation in 10% dimethylsulphoxide containing 200 µM methisazone. This was found to be due to inhibition of phosphodiesterase, the enzyme which is responsible for the degradation of cyclic AMP. Methisazone did not cause any inhibition of phosphodiesterase activity in human lymphocytic cell lines, HeLa cells, or a cell line derived from a mouse lymphoma. The inhibition was therefore specific for normal human lymphocytes, and might be expected to have some influence upon immune responses. An effect of this nature was demonstrated by McNeill et al. (1972), who found that daily intraperitoneal doses of 23 µg caused a reduction in the antibody-forming cell response to injected sheep red cells in mice. The titre of haemolytic antibody was also greatly reduced. The authors concluded that methisazone has a marked immunosuppressive activity. In further work (McNeill, 1972) the effect on colony formation from suspensions of mouse spleen cells was studied. The number of colonies formed during a 7-day period of incubation was reduced when methisazone was incorporated into the medium. The extent of reduction was linearly related to the concentration of the compound, and 50% inhibition occurred at a concentration of 13 µM (3 µg/ml). The inhibitory activity of methisazone was of the same order as that of 6-mercaptopurine (12 µg/ml) and rifampin (10 µg/ml), but did not approach the activity of daunorubicin (6 ng/ml). Exposure of the cells to 10 µg/ml methisazone for 1 hour at 37 °C did not affect their ability to form colonies, and the reduction in colony formation was thus not due to a short-term cytotoxic action. In further experiments the cell suspensions were exposed to methisazone after an initial delay ranging from 3 to 48 hours. The degree of inhibition observed fell off rapidly when the addition was delayed until 24 hours or longer, and methisazone thus exerted its effect at an early stage in colony formation. Methisazone also inhibited the formation of colonies by suspensions of cells from mouse and human bone marrow (McNeill, 1973).

A further biological action of methisazone is its ability to inactivate certain viruses on direct contact with the virion. This was first observed by Levinson et al. (1971) in work done mainly with the 1-ethyl analogue. When Rous sarcoma virus in tissue culture medium was exposed to 40 µM 1-ethylisatin 3-thiosemicarbazone for 15 minutes at 37 °C, more than 90% of the infectivity was lost. The effect was apparently specific, and no inactivation was observed in similar experiments with a representative selection of both RNA and DNA viruses, including herpes, but the compound inactivated herpes when the virus was suspended in phosphate-buffered saline. Methisazone was equally effective, but no inactivation was obtained with the parent compound isatin 3-thiosemicarbazone.

In further work Levinson et al. (1973) observed that Rous sarcoma virus inactivated by methisazone would still attach to chick fibroblasts, and lack of attachment was therefore not the explanation of the loss of transforming activity brought about by methisazone. They also reported that methisazone would inhibit plaque formation by type 1 and type 2 herpes viruses, but plaque formation by polyoma virus was unaffected. In their work on the slow viruses and thiosemicarbazones reported above, Haase and Levinson (1973) found that exposure to 40 µM methisazone over a period of 30 minutes at 37 °C reduced the infectivity of visna virus by 85%.

Inactivation of lymphocytic choriomeningitis virus was reported by Logan et al. (1975), and they also found that methisazone inactivated Parana and Pichinde viruses, which are also members of the arenavirus group.

Antiviral activity

The discovery of methisazone resulted from the original observation of Brownlee and Hamre (1951) that p-aminobenzaldehyde thiosemicarbazone inhibited the multiplication of vaccinia virus. This was the first antiviral compound to be discovered. Thompson et al. (1953) investigated thiosemicarbazones based on other ring systems, and found particularly high activity with isatin 3-thiosemicarbazone. Groups of mice were infected intracerebrally with approximately 1, 10 and 100 LD50 of vaccinia virus and given a diet containing isatin 3-thiosemicarbazone in concentrations of 0·08–0·16%. There was a reduction in mortality of 2 log units in comparison with untreated control groups infected with the same dose levels of virus. These results were confirmed by Bauer (1955), who observed a reduction in mortality of 4 log units in mice given the compound sub-cutaneously in repeated doses of 1·25 and 2·5 mg/kg.

The introduction of continuous cell lines into virology made it possible to examine the antiviral activity of isatin 3-thiosemicarbazone in a more sensitive system. Sheffield *et al.* (1960) demonstrated antiviral activity in tube cultures of HeLa cells infected with decimal dilutions of vaccinia virus and supplied with medium containing isatin 3-thiosemicarbazone in concentrations ranging from 0·4 µM to 40 µM. The infectivity titre as determined by the appearance of cytopathic effect was reduced by 4·7 log units in comparison with infected control tubes incubated in the absence of the compound. This observation did more than confirm the antiviral activity previously found in mice, since it showed that the active agent was the compound itself and not a metabolite produced by enzymatic degrada-tion in the tissues.

Bauer and Sadler (1960 a) examined a number of derivatives of isatin 3-thiosemicarbazone with the object of finding compounds with greater antiviral activity. Most substitutions reduced or abolished activity, but alkylation in the 1- position increased activity, with the 1-methyl and 1-ethyl derivatives respectively having 2 and 3 times the activity of the parent compound. The 1-methyl derivative, methisazone, was selected for further development.

Methisazone inhibits the multiplication of other members of the pox-virus group. In early work with isatin 3-thiosemicarbazone Bock (1957) could find no evidence of antiviral effect in mice infected with ectromelia virus intranasally or by injection into the footpad and treated with the compound subcutaneously in doses ranging from 25 to 250 mg/kg. Bauer and Sadler (1960 a) also found no evidence of activity in mice infected with ectromelia virus intracerebrally, but Sheffield *et al.* (1960) had no difficulty in demonstrating antiviral activity in tissue culture. In tube titrations of ectromelia virus carried out in HeLa cells a concentration of 40 µM reduced the titre of virus by 2·8 log units. It is evident that isatin 3-thio-semicarbazone inhibits the multiplication of ectromelia virus, but the activity is not great enough to be detectable in animal models.

At the time when isatin 3-thiosemicarbazone was considered to be inactive against ectromelia it was considered of interest to investigate its activity against a wider range of pox viruses. In titrations of rabbitpox virus carried out by intracerebral infections with mice it was found that treatment with the compound in doses of 100 mg/kg given twice daily by the subcutaneous route conferred protection against 100 000 LD50 of virus. Rabbitpox was therefore as sensitive as vaccinia (Bauer and Sheffield, 1959). In further work Bauer (1961) obtained dose-response lines based on the mean reciprocal survival times of mice infected intracerebrally with rabbitpox, vaccinia, cowpox and its white variant and treated with isatin 3-thiosemicarbazone in a range of dose levels. The results showed that the sensitivities of these pox viruses covered a wide range, with rabbitpox being the most sensitive, followed by vaccinia and white cowpox with similar sensitivities, and cowpox being the least sensitive, with an ED50 about 1000 times greater than that of rabbitpox.

The most important member of the poxvirus group is smallpox. This exists in two variants, alastrim and variola major. The 1-ethyl derivative of isatin 3-thiosemicarbazone was shown by Bauer and Sadler (1960 b) to be active against alastrim. This virus produces a fatal encephalitis on intracerebral injection in infant mice up to 6 days of age, and subcutaneous treatment with the compound in doses ranging from 10 to 250 mg/kg gave complete protection against infection with 100 LD50 of virus; the 50% protection level was around 0·25 mg/kg. In similar experiments with variola major Bauer et al. (1962) found that subcutaneous injection of isatin 3-thiosemicarbazone in doses of 25 to 100 mg/kg would confer complete protection against intracerebral infection with 1000–10 000 LD50 of virus; the 50% protective dose was around 5 mg/kg. Methisazone was more effective, with a 50% protective dose around 2·5 mg/kg. These findings show that variola major is somewhat less sensitive than alastrim.

Methisazone has antiviral activity against other DNA viruses. Bauer and Sadler (1960 a) found no evidence of activity against herpes in mice infected intracerebrally and treated with 100 mg/kg methisazone by the subcutaneous route, but it is uncertain whether the compound passes the blood–brain barrier. Rapp (1964) found that concentrations of 4·8 μM in the overlay did not reduce plaque formation by herpes, but this cannot be accepted as a negative result since the concentration is very low. Caunt (1967) found that methisazone would reduce the yield of herpes virus in cultures of primary human thyroid cells, with inhibition approaching 100% with a concentration of 20 μM. This was confirmed by Herrmann (1968), who obtained zones of inhibition in plaque inhibition tests carried out in HeLa cells. Methisazone is thus active against herpes, but the activity is too low to be of clinical use.

Rapp (1964) found that methisazone did not inhibit plaque formation by varicella-zoster virus in monolayers of human embryo lung cells, but he was using the same low concentration which was ineffective against herpes virus. In experiments similar to those carried out with herpes virus Caunt (1967) found that methisazone in a concentration of 20 μM reduced the yield of varicella-zoster virus by 90%.

Methisazone is considerably more active against infectious bovine rhinotracheitis, which is also a member of the herpesvirus group. Munro and Sabina (1970) found that methisazone in a concentration of 20 μg/ml (82 μM) reduced the single-cycle yield of the virus in bovine kidney cells by 4–5 log units.

Methisazone will inhibit the multiplication of several types of adenovirus (Bauer and Apostolov, 1966). The production of adenovirus type 11 in HeLa cells was completely inhibited by a concentration of 40 μM, and similar results were obtained with types 3, 7, 9, 14, 16, 17, 21 and 28, and also the simian adenovirus SV15.

Methisazone is also active against a wide range of RNA viruses. No activity was detected in early work, which was carried out by the intracerebral infection of mice. Isatin 3-thiosemicarbazone was reported to be inactive against Rift Valley fever (Minton et al., 1953) and Ilhéus, Wyeomyia, Zika, California, Ntaya, Semliki Forest, dengue 1, Anopheles A and Anopheles B viruses (Bauer and Sadler, 1960). These findings do not exclude activity, however, since it is not known whether the compound can pass the blood–brain barrier.

Activity against rhinoviruses can be readily demonstrated in tissue culture. Bauer et al. (1970) found that methisazone reduced the single-cycle yield of Bunyamwera and Semliki Forest viruses in HeLa cells: the reduction was dose-dependent, and attained 3 log units at a concentration of 20 μM. The inhibition observed was not due to toxicity, since the multiplication of Sindbis virus in the same cell system was unaffected by methisazone in concentrations up to 40 μM. In similar experiments methisazone inhibited the multiplication of echovirus types 7, 11, 12 and 13, reovirus 3, the A/England/66 strain of influenza and the Sendai strain of parainfluenza 1.

Lwoff and Lwoff (1964) found that the single-cycle yield of type 1 poliomyelitis virus was reduced by isatin 3-thiosemicarbazone in a concentration of 40 μM, and Bauer et al. (1970) found that methisazone was active against all three types by the plaque inhibition method.

In an investigation of the effect of various compounds on the multiplication of type A foot-and-mouth disease virus in calf kidney cells Polatnick (1965) found that 100 μg/ml isatin 3-thiosemicarbazone reduced the single-cycle yield by 1·5 log unit; methisazone was more active, and reduced the yield by 4·4 log units in a concentration of 50 μg/ml. Isatin 3-thiosemicarbazone and methisazone have also been reported active against rhinoviruses (Gladych et al., 1969).

The antiviral spectrum of methisazone is given in Table 4.2.

Resistance can be induced in pox viruses by serial passage in the presence of methisazone and its analogues. Appleyard and Way (1966) passaged rabbitpox virus intranasally in mice which were treated with 0·5 or 1 mg of methisazone daily. At each passage the virus was recovered from the lungs and dose-response lines against isatin 3-thiosemicarbazone were obtained by plaque reduction. With successive passages the virus became less sensitive, as indicated by a shift of the dose-response lines towards higher concentrations of the compound. Resistance was also demonstrated in tissue culture by passage in the presence of concentrations of isatin

3-thiosemicarbazone rising from 0·05 to 20 µg/ml. In plaque reduction tests the initial virus was almost completely inhibited by a concentration of 0·1 µg/ml, but after 15 passages a concentration of 10 µg/ml inhibited plaque production by only 50%. The virus had thus become 100 times less sensitive. Katz *et al.* (1973 b) isolated resistant strains of vaccinia virus by passage in the presence of isatin 3-thiosemicarbazone, and also strains which were dependent upon the compound and would not grow in its absence.

Table 4.2   Antiviral spectrum of methisazone (Bauer *et al.*, 1970)

Nucleic acid	Group	Virus
	Poxvirus	Vaccinia Smallpox Cowpox
DNA	Adenovirus	Adenovirus
	Herpesvirus	Varicella-zoster Herpes? Infectious bovine rhinotracheitis
	Picornavirus	Poliomyelitis Echovirus Rhinovirus Foot-and-mouth disease
	Reovirus	Reovirus 3
RNA	Arbovirus	Rift Valley fever Bunyamwera Semliki
	Myxovirus	Influenza A, B
	Paramyxovirus	Parainfluenza 1

Rabbitpox virus also shows another type of resistance which is dependent upon the type of cell in which it is multiplying. Appleyard *et al.* (1965) found that it was less sensitive to isatin 3-thiosemicarbazone in RK13 and L cells than in HeLa cells, and they considered that the compound was 100 times less effective in the first two cell lines.

Mode of action

Although isatin 3-thiosemicarbazone inactivates a number of viruses on direct contact it has no effect on pox viruses, and does not prevent their adsorption to cells (Sheffield *et al.*, 1960). The site of action is thus intracellular. Easterbrook (1962) found that the production of infectious vaccinia virus in KB cells was completely inhibited, but the production of the virus proteins, as observed by immunofluorescence, was unaffected. The virus grew to the normal extent in cultures which had been exposed to isatin 3-thiosemicarbazone for 18 hours and then washed before infection. The antiviral effect was therefore not due to toxicity or the induction of interferon.

In experiments with radioactive thymidine and valine Bach and Magee (1962) showed that isatin 3-thiosemicarbazone did not inhibit the synthesis of DNA and protein in infected HeLa cells. These results are in accordance with the observations of Easterbrook, and indicate that virus components are produced in the presence of isatin 3-thiosemicarbazone but are not assembled into infectious virus particles. Magee and Bach (1965) found that vaccinia virus DNA made in the presence of isatin 3-thiosemicarbazone was incorporated into infective virus particles when the compound was removed, and it was therefore unlikely that any direct interaction between isatin 3-thiosemicarbazone and DNA had occurred.

Appleyard *et al.* (1965) investigated the effect of time of addition of isatin 3-thiosemicarbazone on the yield of rabbitpox virus in HeLa cells. Virus growth was still completely inhibited when treatment was delayed until 3 hours after infection, and was much reduced when the interval was extended to 4 hours. In the absence of treatment infective virus appeared after 5 hours, and it was therefore concluded that isatin 3-thiosemicarbazone exerted its action at a late stage in the growth cycle. However, when the compound was added at the time of infection and removed 2 hours later the final yield of infectious virus was reduced to one-quarter, and isatin 3-thiosemicarbazone must therefore affect an early stage in virus multiplication as well.

The authors also studied the effect of isatin 3-thiosemicarbazone on the formation of virus antigen. Cultures were infected and incubated with the compound for various periods of time, and extracts of the cells were then prepared and examined by immunodiffusion against rabbitpox antiserum. Formation of virus antigen proceeded normally for 4 hours, but after this time the additional lines found with extracts of untreated cultures failed to appear.

In further experiments it was observed that isatin 3-thiosemicarbazone in a concentration of 0·1 µg/ml reduced the 24-hour yield of virus by 95%, but if actinomycin D was present in the culture in a concentration of 0·04 µg/ml the antiviral effect of isatin 3-thiosemicarbazone was almost entirely abolished. The authors interpreted this effect as indicating that DNA-dependent synthesis of RNA must be free to occur if the compound is to manifest its antiviral activity, and they postulated that the presence of isatin 3-thiosemicarbazone brought about the synthesis of a new species of messenger RNA, which was then translated into a protein which was the true antiviral agent.

Woodson and Joklik (1965) applied the methods of molecular biology to a study of the mechanism of action of isatin 3-thiosemicarbazone. In the presence of the compound they found that the amount of virus DNA resistant to DNAse through being coated with virus protein was reduced by 75%. This implies that the synthesis of virus protein is reduced, and is in agreement with the inhibition of late virus antigens observed by Appleyard *et al.* (1965). Reduction in the synthesis of virus proteins was observed in experiments in which the uptake of labelled amino acids was studied. In the presence of 15 µM isatin 3-thiosemicarbazone virus protein was formed at the normal rate for the first 3 hours, but the amount formed then fell to 20% of normal at 4 hours and 10% at 6 hours. The synthesis of virus messenger RNA was unaffected, and the inhibition of protein synthesis must therefore occur at the stage of translation. This was found to proceed normally for the first 4 hours, but the polyribosomes then became unstable and fell away from the messenger RNA strands, which became broken down into shorter lengths. The formation of the late virus proteins was therefore brought to an end. The effect on the messenger RNA could conceivably be due to the action of an abnormal nuclease which is formed as a result of the action of the compound, a step which cannot take place in the presence of actinomycin D.

These results are not in agreement with the observations of Easterbrook (1962), who found that the production of virus proteins was unaffected. Katz *et al.* (1973 a) studied the formation of vaccinia virus polypeptides by polyacrylamide gel electrophoresis, and found that the formation of both early and late virus proteins was unaffected by isatin 3-thiosemicarbazone. They pointed out, however, that the vaccinia virus genome is large enough to code for several hundred polypeptides, and it is therefore possible that the absence of some might not be detectable. They also postulated that the breakdown of polyribosomes and decline in protein synthesis were secondary effects of the compound, resulting from the accumulation of unassembled structural components of the virus.

It is clear from the above that the precise mode of action has not been established, and it is likely that the formation of metal chelates plays some part.

Toxicity

There have been a number of reports on the cytotoxic concentration of methisazone in tissue culture systems. Owing to the low solubility of the compound it is doubtful whether it remains for long in true solution at concentrations above 100 μM (23·4 μg/ml), and some reports have referred to concentrations higher than this. Bauer and Apostolov (1966) found that a concentration of 40 μM was not toxic for HeLa cells, but Munro and Sabina (1970) found that the multiplication of a line of bovine kidney cells was arrested after the third subculture in the presence of a concentration of 20 μg/ml. The compound has some toxic effect on the DNA pathway, since Magee and Bach (1965) found that the incorporation of labelled thymidine by HeLa cells was inhibited to the extent of 65% by a concentration of 50 μg/ml.

The acute oral LD50 of methisazone in mice, rats and rabbits is in excess of 2000 mg/kg. In chronic toxicity studies in which rhesus monkeys were given 250 mg/kg daily by mouth for 1 month there was some evidence of liver damage, but the liver was unaffected at the end of similar treatment in rats and dogs.

Methisazone has an embryotoxic effect in very high doses. In mated rat does given 2000 mg/kg daily by mouth for 12 days there were no implantations in the majority of cases, and animals treated later in pregnancy showed an increase in resorptions in comparison with untreated controls. Similar effects have been noted in rabbits, and fetal malformations are occasionally produced.

The duration of hypnosis after pentobarbitone sodium is greatly prolonged in mice given 25 mg/kg methisazone, probably due to competition for the detoxifying systems in the liver. In rats with unrestricted access to food and water the stomach emptying time is greatly prolonged by the oral administration of 500 mg/kg methisazone. The delay in emptying was much less in fasted rats, and methisazone should therefore be given on an empty stomach.

Metabolism

The metabolism of methisazone has been studied in volunteers and also in contacts of smallpox patients treated for prophylactic reasons. Unaltered drug does not appear in the urine, and five or more metabolites can be detected, including 1-methylisatin, 1-methylisatin 3-thiosemicarbazone, isatin 3-semicarbazone, isatin 3-thiosemicarbazone and 6-hydroxy-1-methylisatin. There is also a possibility that metabolites in the 4- and 7-positions may appear. The hydroxylated derivatives are probably excreted as conjugates with glucuronic and sulphuric acids, and excretion in this form is probably considerably increased in patients in whom the activity of the conjugating systems has been increased by treatment with barbiturates. The metabolites impart a brown colour to the urine, which appears 2–3 hours after taking the compound by mouth and persists for several hours. Methisazone probably interacts with the systems which metabolize ethanol, since it frequently induces a marked intolerance to alcohol. The metabolism of methisazone thus follows a number of pathways, which involve detachment of the methyl group, replacement of sulphur by oxygen in the side-chain and hydroxylation in the aromatic ring. Of the known metabolites only isatin 3-thiosemicarbazone possesses antiviral activity, amounting to one-half that of methisazone.

Clinical pharmacology

Methisazone is poorly absorbed from the gastrointestinal tract. The extent of absorption is greatly influenced by particle size. When given to rats in the form of a suspension in sucrose syrup of micronized particles of mean diameter 3 µm, between 40 and 50% of the dose can be recovered in the faeces. With 100 µm particles the recovery approaches 80% (Axon, 1972).

Methisazone can be detected in the plasma after oral administration, but there is a wide variation in the amounts reported. Turner et al. (1962) found a concentration of 8 µM (1·90 µg/ml) in a child aged 4½ months who had been given 250 mg 6-hourly, a concentration which is known to be inhibitory in tissue culture experiments. Kempe et al. (1965) found that plasma levels attained a maximum between 4 and 8 hours after administration. In an infant with progressive vaccinia a single oral dose of 100 mg/kg gave a peak level of 51·3 µM (12 µg/ml). With 200 mg/kg the maximum level was 85·5 µM (20 µg/ml). In an adult with eczema vaccinatum a single dose of 3 g gave a maximum level of 27·0 µM (6·3 µg/ml). Methisazone had mostly disappeared from the plasma after 12 hours, and repeated administration did not lead to any cumulative rise in plasma concentration. Much lower values were found by Gomez and Sandeman (1966) in one normal subject and six patients with malignant conditions treated with methisazone in doses of 40 and 80 mg/kg. Measurable levels ranging from 1·3 µM (0·3 µg/ml) to 3·4 µM (0·8 µg/ml) were found in three and trace amounts in two; in one patient methisazone could not be detected in the serum. No information is available on the concentrations of methisazone attained in the tissues.

Assay method

Methisazone readily dissolves in alkali to form a yellow solution with an absorption maximum at 400 nm. In the method developed by Turner et al. (1962) 1 ml of heparinized blood is acidified with 0·1 ml of 2 N hydrochloric acid and extracted into 1 ml of toluene. After centrifugation for 20 minutes at 2500 g 0·5 ml of the toluene layer is basified with 0·1 ml of 0·05 M methylbenzethonium hydroxide in ethanol and the optical density is read at a wavelength of 400 nm. The concentration of methisazone is then determined from a standard line obtained by the treatment of solutions of known concentration. The determination may be carried out more conveniently by extraction of the acidified sample into benzene and re-extraction of the benzene solution into 1 N sodium hydroxide solution. Readings should be made without delay, as the colour of the alkaline solution fades rapidly.

Clinical use

Methisazone is indicated for the prophylaxis of smallpox in contacts and for the treatment of infections of the skin with vaccinia virus. It is also used to cover primary smallpox vaccination when it must be carried out in the presence of contraindications, and it has been reported to have some effect in the treatment of varicellar pneumonia.

Contraindications

Methisazone is contraindicated in patients with liver dysfunction unless they are contacts of smallpox or are suffering from a severe complication of smallpox vaccination. The decision whether or not to use it in pregnancy should be weighed against the known severe course of smallpox in pregnant women and the risk to the fetus.

Preparations

Marboran suspension (Wellcome Foundation Limited). Sachets containing 15 ml of a 20% suspension of Marboran in sucrose syrup (3 g per sachet).

GENERAL READING

Hoffmann, C. E. (1973). Amantadine HCl and related compounds. In: W. A. Carter (ed.) *Selective Inhibitors of Viral Functions*, pp. 199–211, (Cleveland Ohio: Chemical Rubber Co. Press)

Bauer, D. J. (1972). Thiosemicarbazones. In: D. J. Bauer (ed.) *Chemotherapy of Virus Diseases*, vol. 1, 35–113. (Oxford: Pergamon Press)

Chapter 5

Chemotherapy of herpesvirus infections—I

Herpes and B virus

The herpesviruses contain two genera, Herpesvirus and Cytomegalovirus. A simplified classification is shown in Table 5.1. The genus Herpesvirus consists of two recognized subgroups. Subgroup A contains three viruses which infect man: herpes (simplex), B virus and pseudorabies. Subgroup B contains only varicella-zoster. Apart from these there are a number of viruses of animals which await classification. The genus Cytomegalovirus

Table 5.1 Classification of the herpesviruses

Genus	Subgroup	Virus
Herpesvirus	A	Herpes (simplex) B virus Pseudorabies Related viruses of domestic animals
	B	Varicella-zoster
	Unclassified	Epstein-Barr virus Avian and amphibian viruses of herpes type
Cytomegalovirus	—	Cytomegalic inclusion disease Salivary gland viruses of rodents and other animals

contains one virus infecting man, that of cytomegalic inclusion disease, and a number of related viruses mainly affecting rodents.

Most virus infections are acute and self-limiting, and are brought to an end by the development of immunity or death of the host. Second attacks by the same agent are rare or non-existent, and further attacks of diseases such as influenza or the common cold are due to new infections with variants of the virus which are antigenically distinct and to which the patient has thus no immunity. With the viruses of the herpes group the situation is quite different. On first experience with the virus the patient undergoes a primary infection. On recovery, the virus may persist in the dorsal root ganglia which innervate the site of infection. The persistent infection is asymptomatic, but it may become reactivated, with the formation of virions which pass down the axons to the area of the skin or cornea innervated by them. A recurrent infection then develops locally which is often of similar type to the initial infection. The mechanisms of latency and reactivation will be dealt with more fully in the sections on herpes and varicella-zoster.

HERPESVIRUS INFECTIONS OF SUBGROUP A

Herpes (simplex)

In this work the term herpes will be used to refer to the virus commonly termed herpes simplex virus, since herpes simplex is a skin disease and only one particular manifestation of infection with the virus, and it is therefore illogical to refer to herpes simplex in connexion with other manifestations of infection with the same virus, such as encephalities and keratitis. The virus is also referred to as herpesvirus hominis; this is correct, since it is the approved binomial, but it is rather cumbersome to use in the context of clinical work.

Strains of herpes virus fall into two types, which can be readily distinguished by their laboratory characteristics. Strains of type 1 are responsible for extragenital infections, and type 2 strains cause the majority of genital infections, although type 1 strains are isolated not infrequently.

Primary infection: localized

Antibody studies indicate that the primary infection usually occurs early in life. It is usually localized, with lesions at the portal of entry and enlargement of the regional lymph nodes. In many instances the infection is trivial and may pass unnoticed.

Gingivostomatitis

The most frequent site of infection is the mouth, and the condition produced may range from a small ulcer on the tongue to extensive gingivostomatitis, with pyrexia and enlargement of lymph nodes, leading to the development of painful ulcers which heal slowly over the period of a week or so. The condition is particularly severe in patients who are treated with steroids or immunosuppressants. Isolated lesions may be treated with 5% idoxuridine in Orabase, which adheres to the mucosa (Najjar *et al.*, 1969). A 40% solution of idoxuridine in dimethylsulphoxide used as a paint is also effective, but the taste of the solvent is objectionable to some patients. Severe cases should be treated with cytarabine in a dose of 3 mg/kg, followed by 2 mg/kg daily for up to 4 days. This treatment is unlikely to cause depression of the bone marrow (Juel-Jensen and MacCallum, 1972).

Herpetic whitlow

Primary herpetic infections of the normal skin may be trivial or escape notice, except for the form known as herpetic whitlow. This occurs in young adults, particularly medical students, doctors, dentists and nurses, who may be called upon to examine the mouth of a person suffering from oral herpes. A series of 54 cases among hospital nurses has been published by Stern *et al.* (1959). The infection begins with intense itching and pain and the formation of vesicles, which coalesce and enlarge. There is often fever and enlargement of lymph nodes, and the infection may result in extensive local destruction of tissue. The condition usually progresses for 10 days, with continuing pain; it then begins to regress, and healing may be complete by 3 weeks from the time of onset. Rapid improvement occurs on treatment with 40% idoxuridine in dimethylsulphoxide (Juel-Jensen, 1971). The lesion is covered with lint which is kept continuously moist with the solution. In the series of eight cases reported the mean duration of pain was reduced to 1·6 days and healing was complete by 2 weeks. Treatment should not be prolonged unnecessarily, or the continuous application of dimethylsulphoxide will lead to maceration of the skin and necrosis. A typical response to treatment is illustrated in Figure 5.1.

Eczema herpeticum

This condition was first described as Kaposi's varicelliform eruption, a term which also covers a similar condition caused by infection with vaccinia virus. It occurs in persons with atopic eczema, and takes the form of a generalized vesicular eruption, with enlargement of lymph nodes, and fever. It occurs most frequently in children. It responds rapidly to treatment with cytarabine given intravenously in a dose of 2 mg/kg daily on 5 successive days (Juel-Jensen and MacCallum, 1972).

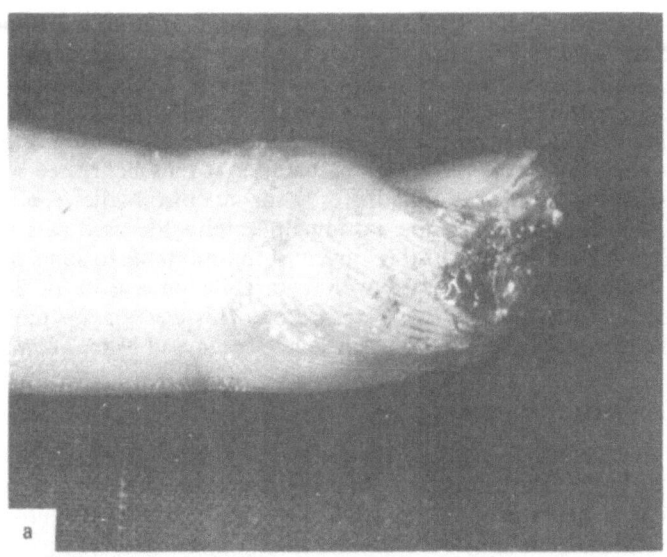

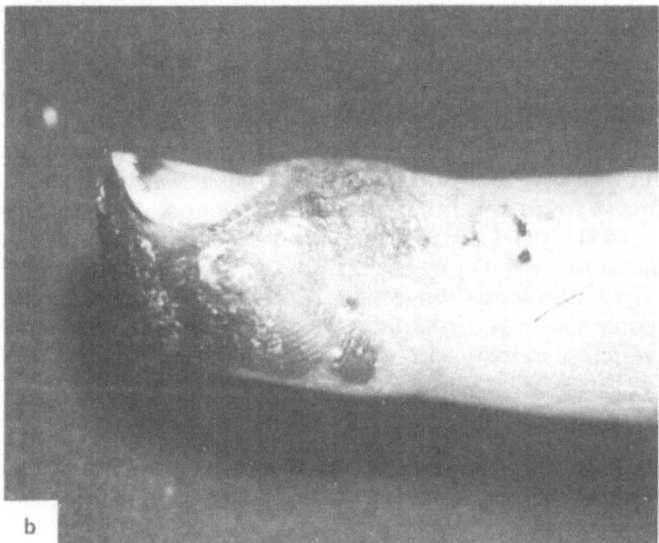

Figure 5.1 Treatment of herpetic whitlow with 40 % idoxuridine in dimethylsulphoxide. (a) Initial state. (b) After surgical incision. (c) After
treatment with idoxuridine. (d) Final state

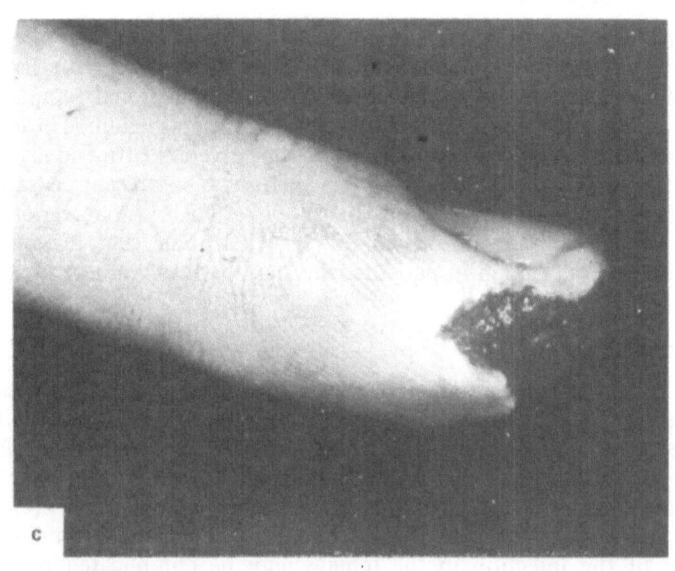

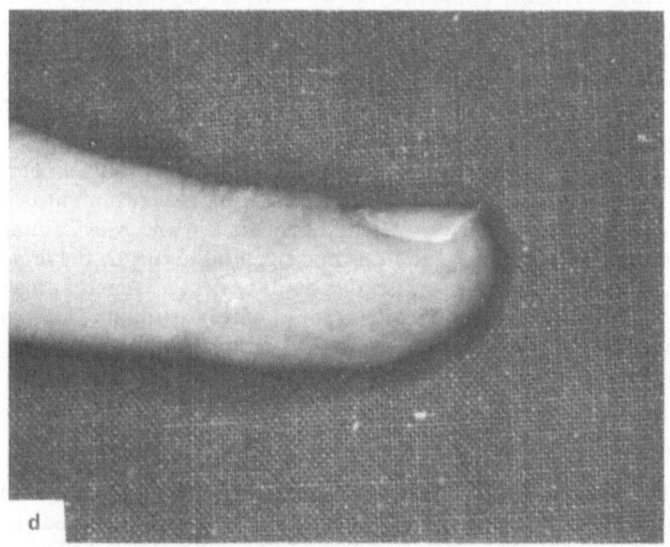

Follicular conjunctivitis

When the primary infection occurs in the eyes it gives rise to follicular keratitis with regional lymphadenitis. There may be vesicles on the lids and coarse opacities in the corneal epithelium, with regional lymphadenitis. Punctate opacities are frequently observed in the epithelium. The condition should be treated with idoxuridine eye drops given five times daily for 7–10 days, or until any corneal opacities which may be present cease to stain with bengal rose (Patterson and Jones, 1967). There are no reports as yet of treatment with trifluorothymidine or vidarabine, but it would seem reasonable to use them in patients who may be intolerant of idoxuridine.

Genital herpes

Primary herpes can occur as a venereal infection, with vesicular lesions in the genital region, usually caused by type 2 virus. The condition may be treated by the local application of 20% or 40% idoxuridine in dimethyl-sulphoxide. Vidarabine was found to be ineffective in a trial in which a 3% ointment was used topically in males with active genital herpes (Goodman et al., 1975). There are no reports of the use of trifluorothymidine. The course of the infection in the female may be complicated by a severe necrotizing cervicitis, apparently first described by Willcox (1968). In a series of six cases the main symptoms were vaginal discharge, dysuria, local soreness and pain in the lower abdomen and right iliac fossa. In all patients there was gross inflammation of the cervix, from which pieces of dead tissue could be removed. The cervix bled easily and was painful to touch in four patients. Healing was slow but was usually complete by the third week. No specific treatment has been reported. A solution of idoxuridine would not reach the affected area, but some success might be obtainable if it were formulated as a pessary. Juel-Jensen and MacCallum (1972) have pointed out that the pain in primary genital herpes in the female may be so severe that examination is not possible without general anaesthesia. They described two cases in which a rapid cure was obtained with cytarabine given in a single intravenous dose of 4 mg/kg followed by 3 mg/kg on the 4 following days.

Encephalitis

Herpetic encephalitis is a particular manifestation of localized primary herpes in which the brain is the organ predominantly affected. It occurs mainly in infants, often in the first days of life, and is due to infection with a type 2 strain acquired from the mother in the perinatal period. In a series of 148 cases of neonatal herpes infection analysed by Nahmias et al. (1970), the brain was the only or main organ affected in 25; 11 cases were fatal, and of the survivors 12 were left with neurological sequelae. The onset was at birth or up to 21 days later, with a mean value of 6 days. The course was severe, with rapid deterioration and death in the fatal cases occurred after 3 hours to 6 days. The symptoms were frequently unspecific at first, with poor feeding and weight gain, vomiting, diarrhoea, respiratory difficulty and sometimes fever. The most frequent signs of involvement of the brain were convulsions, opisthotonus, bulging fontanelle, disturbances in muscle tone and coma. There was difficulty in establishing the diagnosis unless lesions could be seen at external sites.

Herpetic encephalitis may also occur in adults who have not hitherto experienced a primary infection, although it is possible that some cases may be recurrent infections. It is usually caused by type 1 virus. The symptoms and clinical course have been described by Illis and Gostling (1972) and Zischka-Konorsa et al. (1965). Three clinical types have been described; multifocal disseminated meningoencephalitis, a massive necrotizing and haemorrhagic inflammation of the brain with acute or protracted course typically with unilateral involvement of the temporal lobe and limbic system, and a subacute or chronic sclerosing form. The variation in clinical type is probably due to host factors rather than variation in the properties of the virus. The onset is very variable. The condition may mimic a space-occupying lesion, or begin as a change in personality with difficulty in speech leading to a febrile illness with convulsions. It may also take the form of acute encephalitis or pursue a biphasic course. Focal signs are common, such as hemiparesis, weakness of the facial muscles and cranial nerve palsies. The mortality as reported in the literature ranges from 13% to 100%, with a mean value of 70%.

The treatment of herpetic encephalitis is very unsatisfactory, in striking contrast to the generally favourable results of treatment of herpetic infections at other sites. The first report of specific treatment is that of Breeden et al. (1966), who used idoxuridine in the treatment of a man aged 34. Treatment was begun on the 22nd day of illness in a total dose of 520 mg/kg given over 7 days by intravenous infusion. The patient showed improvement and recovered, and it was considered that idoxuridine might have had a favourable effect. An apparently favourable result was also obtained by Evans et al. (1967) in an 8-year old girl who was given a total dose of 500 mg/kg in the form of intravenous infusions of 1·5 g over a period of 8 hours on 5 alternate days. Marshall (1967) reported a rapid recovery in a 13-year old boy who had undergone external decompression without relief of symptoms.

The administration of idoxuridine has since become standard practice in the treatment of herpetic encephalitis.

An analysis of the cases reported in the literature is given in Table 5.2, which is based on a compilation of Illis and Merry (1972) with subsequent additions. Treatment with idoxuridine appears to reduce the mortality from

Table 5.2 Summary of published cases of herpetic encephalitis treated with idoxuridine

Reference	Age	Duration of illness before treatment (days)	Total dose	Duration (days)	Outcome	Sequelae
Breeden et al. (1966)	34 years	22	520 mg/kg	8	Recovered	Minimal
Evans et al. (1967)	8 years	55	500 mg/kg	8*	Recovered	Severe
Marshall (1967)	13 years	15	200 mg/kg	3	Recovered	Minor
Buckley and MacCallum (1967)	41 years	11	1 g†	—	Recovered	Severe
Bellanti et al. (1968)	11 months	14	310 mg/kg	5	Recovered	Minimal
Rappel and Brihaye (1969)	58 years	16	500 mg/kg	6	Died	—
	12 years	15	324 mg/kg	6	Recovered	Minimal
	56 years	27	500 mg/kg	7	Recovered	Severe
	45 years	18	500 mg/kg	4	Recovered	Minimal
	65 years	9	500 mg/kg	4	Died	—
Duffy (1969)	43 years	—	400 mg/kg	—	Recovered	—
	4 months	—	'similar'	—	Died	—
	Not given	—	'similar'	—	Died	—
Dayan and Lewis (1969)	60 years	—	30 g‡	8	Died	—
Goldman et al. (1970)	20 years	12	1000 mg/kg	5	Recovered	Minor
Meyer et al. (1970)	57 years	13	24 g	5	Died	—
	24 years	13	30 g	5	Recovered	None
	22 years	14	30 g	5	Recovered	None
	10 years	8	15 g	5	Recovered	None
	12 years	3	30 g	5	Recovered	None
	12 years	6	6 g	1	Died	—

* Given on alternate days   † Intracarotid injection   ‡ Given into superior vena cava

Table 5.2—continued

Reference	Age	Duration of illness before treatment (days)	Total dose	Duration (days)	Outcome	Sequelae
Silk and Roome (1970)	6 years	8	550 mg/kg	8	Recovered	Severe
Charnock and Cramblett (1970)	19 days	15	410 mg/kg	7	Recovered	Severe
Tomlinson and MacCallum (1970)	47 years	3	200 mg/kg	2	Died	—
	—	5	400 mg/kg	4	Died	—
	—	8	400 mg/kg	4	Died	—
Nolan et al. (1970)	57 years	—	15 g	3	Died	—
	62 years	—	9 g	3	Died	—
	—	—	430 mg/kg	5	Died	—
	—	—	430 mg/kg	5	Recovered	None
	—	—	430 mg/kg	5	Recovered	None
	—	—	430 mg/kg	5	Recovered	None
	—	—	430 mg/kg	5	Recovered	None
Fishman et al. (1971)	12½ years	5	400 mg/kg	4	Died	—
Johnson et al. (1972)	47 years	10	15 g	5	Died	—
Boston Inter-hospital (1975)	21 years	5	500 mg/kg	5	Died	—
	70 years	4	500 mg/kg	5	Died	—
	52 years	12	500 mg/kg	5	Recovered	Severe
	32 years	5	500 mg/kg	5	Died	—
	17 years	10	500 mg/qg	5	Died	—
	50 years	10	500 mg/kg	5	Died	—
	45 years	4	500 mg/kg	5	Recovered	Minimal
	68 years	11	500 mg/kg	5	Died	—

70% to 30%, but on recovery the patients are often left with severe neurological sequelae. There is a better chance of success if treatment is begun early in the course of the disease, but some effect can be obtained in patients treated several weeks after onset. Adverse side effects such as depression of the bone marrow, alopecia, glossitis and stomatitis are frequent, but do not contribute to the mortality.

Illis and Merry concluded that idoxuridine was the treatment of choice in herpetic encephalitis. Other observers have not gained such a favourable impression. Thus, Fishman *et al.* (1971) reported that idoxuridine failed to modify the course of illness in a girl aged 12½ years who was treated on the 5th day of illness. It was given in a daily dose of 100 mg/kg in five divided doses each given intravenously over a period of 2 hours. There was no improvement and death occurred on the 10th day of illness, after a total dose of 400 mg/kg had been administered.

As the result of growing doubts on the efficacy of idoxuridine in the treatment of herpetic encephalitis a multicentre placebo-controlled double-blind trial was set up in 1971–72 (Boston Interhospital Virus Study Group, 1975). The treatment schedule selected was 100 mg/kg daily by intravenous injection for 5 days. There was no response to treatment and death occurred in five of six patients treated with idoxuridine. There was also a degree of suppression of bone marrow function which was considered unacceptable, and the studies were terminated prematurely in 1973.

The general view now seems to be that idoxuridine is no longer indicated in the treatment of herpetic encephalitis. Juel-Jensen and MacCallum (1972) reported a case in which treatment with cytarabine appeared to be effective. It was given in a single rapid intravenous injection in a dose of 10 mg/kg, followed by 8 mg/kg daily for 4 days and 6·5 mg/kg daily for 2 days. Farris and Blaw (1972) observed a rapid improvement leading to almost complete recovery in a girl aged 13¾ years, who was given 4 mg/kg cytarabine in a single rapid intravenous dose, followed by 6 mg/kg on the next day and 8 mg/kg on the following 3 days. Favourable results were also obtained in four of six cases by Chow *et al.* (1973), and in two of three cases by Lagerkvist and Ekelund (1975). A brief analysis of the cases reported in the literature is given in Table 5.3.

Cytarabine has the advantage over idoxuridine of being much more soluble. It can thus be given in single rapid intravenous injections. Its place in the treatment of herpetic encephalitis needs further evaluation, and a multicentre double-blind trial is being carried out at present in Great Britain, and the results are expected in 1977.

Vidarabine is also being used, but reports of its effectiveness have not yet appeared. It has the advantage of being able to cross the blood–brain barrier, although its intrinsic antiviral activity is much lower than that of the other compounds.

Table 5.3 Summary of published cases of herpetic encephalitis treated with cytarabine

Reference	Age	Duration of illness before treatment (days)	Dose	Duration (days)	Outcome	Sequelae
Juel-Jensen and MacCallum (1972)	21 years	10	10 mg/kg	1	Recovered	Minimal
			8 mg/kg	4		
			6·5 mg/kg	2		
Farris and Blaw (1972)	14 years	12	4 mg/kg	1	Recovered	Aphasia
			6 mg/kg	1		
			8 mg/kg	4		
Chow et al. (1973)	14 days	—	30 mg/m²	5	Recovered	None
			10 mg/m²*	3		
	14 days	—	30 mg/m²	5	Recovered	Severe
			10 mg/m²*	3		
	34 years	—	35 mg/m²	5	Recovered	Minimal
			10 mg/m²*	3		
	29 years	—	160 mg/m²	1	†Recovered	None
			60 mg/m²	1		
			40 mg/m²	2		
	39 years	—	65 mg/m²	1	Died	—
			40 mg/m²	4		
	55 years	—	60 mg/m²	4	Died	—
			30 mg/m²	1		
			10 mg/m²*	4		
Lagerkvist and Ekelund (1975)	6 weeks	13	3 mg/kg	5	Recovered	None
	10 weeks	16	3 mg/kg	5	Died	—
	6 weeks	14	3 mg/kg	5	Recovered	None

* Given intrathecally   † Also treated with idoxuridine

Primary infection: generalized

Disseminated herpes of neonates

While neonatal infection with herpes virus often takes the form of encephalitis it is often generalized, with involvement of the liver, adrenals, trachea, lungs, oesophagus, stomach, kidneys, spleen, pancreas, heart and bone marrow. Encephalitis may also be present as part of the general involvement. The course and outcome are the same as those given in the section on encephalitis. In a series of 98 cases reported by Nahmias *et al.* (1970) the central nervous system was involved in 40. Only four patients recovered.

The condition has been treated with idoxuridine. The details of treatment and results obtained in ten cases are shown in Table 5.4, which is based on an analysis given by Hanshaw (1973). In the case described by Golden *et al.* (1969) improvement set in within 24 hours of beginning treatment, but the patient was nevertheless left with severe sequelae.

John and Tobin (1973) reported two cases treated with cytarabine given intravenously in daily doses of 5 mg/kg for 5 days. One recovered without sequelae, but the other died with bacterial bronchopneumonia. Brown and Bower (1975) obtained recovery with sequelae in an infant which developed disseminated herpes with associated encephalitis on the 12th day of life. The dosage regime used was not stated explicitly, but cytarabine was apparently given in an initial dose of 10 mg/kg, followed by 8 mg/kg daily for 4 days and 6·5 mg/kg daily for 8 days.

Ch'ien *et al.* (1975 a) used vidarabine in the treatment of 13 cases of neonatal herpes, including eight cases of disseminated infection. It was given by intravenous drip over a period of 12 hours in a daily dose of 10–20 mg/kg for 10–15 days. In four patients treatment was begun early in the course of the infection, and recovery was complete in all cases without sequelae. In the other four patients treatment was begun much later and the infection terminated in death.

It may be concluded that the results of treatment of disseminated herpes are far from satisfactory. The mortality appears to be lowered by treatment and it is desirable to begin treatment as early in the course of the disease as possible. Antiviral agents will naturally not reverse damage to organs which has already occurred.

Generalized herpes

When primary herpes infection is delayed until early adult life it may take a generalized form, with a skin eruption, fever, and enlargement of lymph glands and spleen. Juel-Jensen and MacCallum (1972) have described two cases treated with cytarabine. Low doses were used in the first case in order to avoid immunosuppression, with 0·3 mg/kg given intravenously on 5

Table 5.4 Summary of published cases of disseminated herpes treated with idoxuridine

Reference	Age	Duration of illness before treatment (days)	Dose	Duration (days)	Outcome	Sequelae
Partridge and Mills (1968)	—	—	580 mg/kg	—	Died	—
Golden et al. (1969)	4 days	8	500 mg/kg	5*	Recovered	Severe
Tuffli and Nahmias (1969)	7 days	60	200 mg/kg	4	Recovered	Severe
	14 days	16	250 mg/kg	5	Recovered	None
Charnock and Cramblett (1970)	11 days	7	410 mg/kg	7	Recovered	Moderate
Pettay et al. (1972)	—	—	420 mg/kg	—	Recovered	—
	—	—	275 mg/kg	—	Died	
Hanshaw (1973)	—	—	280 mg/kg	—	Recovered	Moderate
	—	—	500 mg/kg	—	Recovered	Severe
	—	—	500 mg/kg	—	Recovered	None

* Course repeated after recurrence of infection

successive days. In the second patient 4 mg/kg was given on the 1st day, followed by 2 mg/kg on the following 4 days. In both cases the condition improved rapidly after beginning treatment and recovery was complete.

No reports on the use of vidarabine have yet appeared, but it would seem reasonable to use it in the dose used by Ch'ien *et al.* (1975 a) in the treatment of neonatal herpes.

Latency and recurrence

After recovery from the primary infection herpes virus may persist in the body and subsequently give rise to recurrent attacks in spite of the continued presence of neutralizing antibody. There is much evidence that during the latent period which elapses between primary infection and recurrence the virus is present in some form or other in the neurones of the dorsal root ganglia supplying the segment involved. Thus, Stevens and Cook (1973 a) found that herpes virus could be found in explant cultures of the spinal ganglia of mice which had survived an infection caused by injecting the virus into the footpad. The virus could not be found in the homogenized ganglia and could not be seen in the neurones on electron microscopy. No virus antigens could be detected by immunofluorescence, but after cultivation of the explants for 7 days infective virus reappeared and could be detected by infecting cultures of RK-13 cells. In further work (1973 b), explants were examined in the electron microscope after some days of cultivation, and virus particles could be seen in the nuclei of both neurones and Schwann cells, and also lying in an extracellular position. By co-cultivation of explants with RK-13 cells Knotts *et al.* (1973) found that herpes virus persisted in the trigeminal ganglia for at least 12 months in rabbits which had developed encephalitis with subsequent recovery after inoculation of the cornea.

Herpes virus has also been isolated from the trigeminal ganglion in man. Bastian *et al.* (1972) found it in autopsy material from the ganglia of one of 22 patients with no clinical evidence of active herpes infection at the time of death. The virus was detected by co-cultivation of explants with VERO cells. In a similar investigation Baringer and Swoveland (1973) found herpes virus in the trigeminal ganglia of six of seven unselected cadavers undergoing autopsy less than 12 hours after death. No virus could be isolated from the roots or the branches of the trigeminal nerve. Only two patients had had an illness suggesting a herpes infection.

The trigeminal ganglia are not the only ones affected. Baringer (1974) removed various regional ganglia from 26 unselected cadavers and co-cultivated explants with human embryo fibroblast cells. Type 2 herpes virus was isolated from the 2nd, 3rd and 4th sacral ganglia in three instances, although there had been no history of recurrent genital herpes.

During latency herpes virus can also be detected in extraneural tissues.

Kaufman *et al.* (1967) infected 15 rabbits on the cornea with a strain of herpes virus which caused recurrent ulceration. After the initial infection had subsided swabs were taken from the cornea at intervals up to 95 days after infection and inoculated into tissue cultures. All the rabbits underwent at least three episodes during which the virus could be isolated from the corneal swabs and ten had at least one recurrent ulcer. The virus was detected in the lacrimal gland before the ulcers appeared, and the authors suggest that a generalized infection of the periocular structures was taking place, thus providing the virus for initiating the recurrent ulcer.

Swabs were also taken daily over a period of 20 days from 11 persons who had no history of herpes infection. The virus was detected in the saliva and eyes in four instances, in the absence of any signs of infection. In a similar investigation of 24 persons who suffered from recurrent herpes infection 13 isolations were made, six from saliva and seven from the eyes. Herpes antigen was found by means of immunofluorescence in sections of the lacrimal gland and cornea in three of four patients with active disease and in four of seven patients during a phase of clinical inactivity (Kaufman *et al.*, 1968). On the basis of these observations the authors suggested that herpes virus did not enter a true state of latency but persisted as a chronic infection of the ocular tissues. This hypothesis should be assessed in the light of the subsequent observations of Stevens *et al.* (1972) that the virus could be detected in the trigeminal ganglia. However, Dawson *et al.* (1968) were able to detect herpes virus in the cornea by electron microscopy. Specimens of cornea were obtained from patients who were undergoing keratoplasty for the sequelae of herpetic keratitis. Particles resembling herpes virus were seen in material from five patients. They were present in the nuclei and cytoplasm of the stromal cells, and they could also be seen in an extracellular position.

Herpes may also remain in a latent state in the cells of the cervix, and there is evidence to suggest that carcinoma of the cervix is a late sequel of herpetic cervicitis. Aurelian (1973) obtained a cell line from a cervical tumour which underwent spontaneous degeneration after 12–18 passages. Type 2 herpes virus could be isolated from the degenerating cultures by the infection of HEp-2 cells, but virus could not be isolated before the onset of degeneration. These observations could be explained if the virus genome became integrated into the cell DNA, and there is evidence that this can happen for at least that part of the herpes genome which codes for the synthesis of thymidine kinase (Munyon *et al.*, 1971).

Reactivation can occur in response to a number of non-specific stimuli, the most familiar being exposure to sunlight, and fever, particularly in pneumococcal pneumonia. A recurrent attack of facial herpes may occur after section of the posterior root of the trigeminal ganglia in patients with trigeminal neuralgia. General reactivation frequently occurs in transplant patients. Lopez *et al.* (1974) carried out a systematic screening of 61 immunosuppressed renal transplant recipients by the examination of urine, sputum and stool specimens for the presence of virus. Herpes virus was detected in 15 patients, and the finding of the virus usually coincided with the onset of fever, leukopenia and graft rejection. The patients continued to excrete the virus after recovery and the development of antibody.

Reactivation occurs in spite of the presence of neturalizing antibody. Numerous attempts to detect virus in the skin at the site of recurrence have been unsuccessful, and since the virus can be detected in the regional ganglia it is reasonable to suppose that it travels down some component of the nerve trunk and gives rise to an infection in the area of innervation. During this process it is presumably not in contact with antibody.

Recurrent infection: cutaneous herpes (herpes simplex, herpes febrilis, herpes labialis)

When the primary infection is in the oral cavity the recurrent infection commonly takes the form of a vesicular eruption on the mucocutaneous junction of the lips, and the angle of the nose may also be affected. The course has been described by Juel-Jensen and MacCallum (1972). For some hours before the appearance of lesions there is tingling and itching at the site, a symptom which suggests that the sensory nerves are somehow involved. A cluster of raised papules then appears, developing into vesicles within 24 hours, with a surrounding area of erythema. The vesicles burst and heal by scab formation. The duration of the eruption ranges from 2 days to 3 weeks, with an average period of 9–10 days.

The association between section of the sensory root of the trigeminal ganglion and the appearance of herpes was probably first reported by Cushing in 1904. Carlton and Kilbourne (1952) observed this phenomenon in 16 of 17 patients who underwent the operation for the treatment of trigeminal neuralgia. The lesions appeared 48–96 hours after the operation on the site of the sectioned root and persisted for 5 days. They usually occurred on the lips at the mucocutaneous junction, and also on the hard palate and cheek, and thus occurred within the area of distribution of the 2nd and 3rd divisions. The eyes were not involved.

Cutaneous herpes is treated with idoxuridine. It was first used by Hall-Smith et al. (1962) in the form of 0·1% solution and 0·5% ointment intended for ophthalmic use. The solution was usually applied hourly during waking hours and the ointment was applied four times a day; the lesions were on the upper lip, eyelid, nose and cheek, and 13 patients were treated in all. Relief of symptoms was noted shortly after beginning treatment and resolution was complete in 3 or 4 days in all patients except one, in whom the course of treatment was thought to be incomplete. A similar favourable result was claimed by Jackson in 22 patients suffering from recurrent herpes of the face and external genitalia, who were treated with idoxuridine in the same formulations. Ophthalmic preparations are not designed to facilitate the penetration of drugs into the skin, and it is therefore not surprising that the reported therapeutic effect of idoxuridine could not be confirmed by Juel-Jensen and MacCallum (1964) in a double-blind trial of the same ointment used by Hall-Smith *et al.*, carried out on 18 patients with recurrent herpes labialis. Eleven patients were treated with idoxuridine and seven received the ointment base as a placebo. The mean duration of the attacks was 8·9 days and 8·6 days respectively and no therapeutic effect was therefore obtained.

With the object of obtaining better penetration of the drug into the skin, MacCallum and Juel-Jensen (1966) used a 5% solution of idoxuridine in dimethylsulphoxide, a solvent which will dissolve idoxuridine to the extent of at least 40%. A trial was carried out on 16 patients, five of whom were treated later for a further attack. Ten attacks were treated with the solution applied with a paint brush three times daily for 3 days, and 11 were treated with the solvent alone as a control group. The mean duration to complete healing was 3·50 days in the treatment group and 5·45 days in the control group. The mean duration expected on the basis of earlier attacks was 9·55 days, and treatment with idoxuridine thus achieved a reduction in duration of 64%. Dimethylsulphoxide alone gave a 43% reduction, and the solvent evidently had some effect on its own.

Trials carried out in guinea-pigs indicate that higher concentrations are desirable for achieving a satisfactory therapeutic effect (Collins and Bauer, 1977). The result of a typical experiment is shown in Figure 5.2. Here both flanks of a guinea-pig were epilated and inoculated at several sites with herpes virus by dermal scarification. The lesions were allowed to develop for 36 hours; those on one flank were then painted with 10% idoxuridine in dimethylsulphoxide twice daily for 4 days. The illustrations show the appearances on the 5th day of infection. On the treated side healing is complete, whereas on the untreated side the lesions have reached the pustular stage and healing will not be complete for another 6–7 days. A similar good therapeutic result can be obtained with a 20% solution, but with a 5% solution the effect is considerably less marked.

Dimethylsulphoxide is not licensed for use in the USA, and Najjar *et al.* (1969) have therefore investigated the effect of 10% idoxuridine in Plastibase, an ointment base consisting of 5% polythene in liquid paraffin. Sixteen patients with herpetic lesions in the mouth were treated, and the lesions resolved in 3 days. Oral lesions responded equally well to the application of 10% idoxuridine in Orabase, an ointment base containing equal parts of sodium carboxymethylcellulose, pectin and gelatin to a total of 50% in Plastibase.

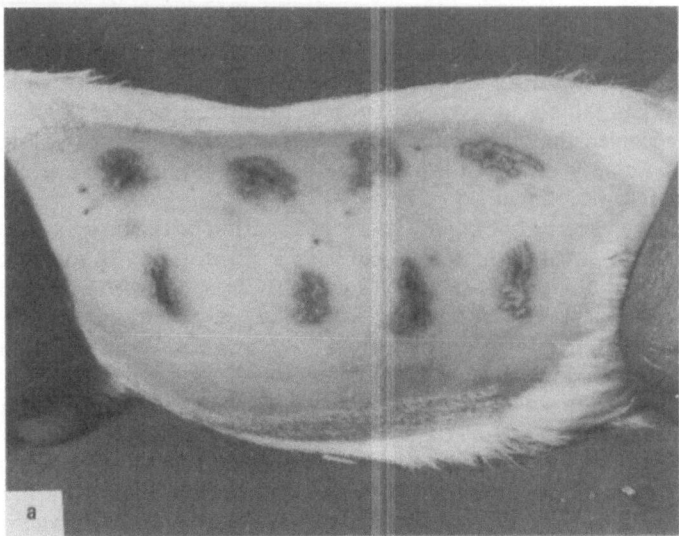

Figure 5.2 Treatment of cutaneous herpes with 20% idoxuridine in dimethylsulphoxide. Type 1 herpes virus was inoculated by dermal scarification into both flanks of a guinea-pig. After 36 hours the
developing lesions on one flank were painted twice daily with
idoxuridine solution. The illustrations show the appearances 4 days
later. (a) Untreated side. The lesions have advanced to the stage
of pustulation and scabbing. (b) Treated side. The lesions have
regressed completely, and the skin shows only the residual signs
of trauma from scarification

Genital herpes

Recurrences of genital herpes are generally much less severe than the
primary infection, and respond well to local treatment with 20% idoxuridine in dimethylsulphoxide.

There is one report on the use of vidarabine in herpes genitalis. Goodman
et al. (1975) investigated the effect of 3% vidarabine ointment in comparison with a placebo in a double-blind trial carried out on 32 male patients.
The ointment was made up in a base consisting of 60% petrolatum and
40% mineral oil (both USP), and was applied to the lesions four times a day
for 7 days. The ointment base alone was used as the placebo. No differences
in the clinical course could be detected in the two groups, and the authors
concluded that vidarabine was ineffective in genital herpes. Since vidarabine
is so much less active against herpes than idoxuridine it would seem neces-

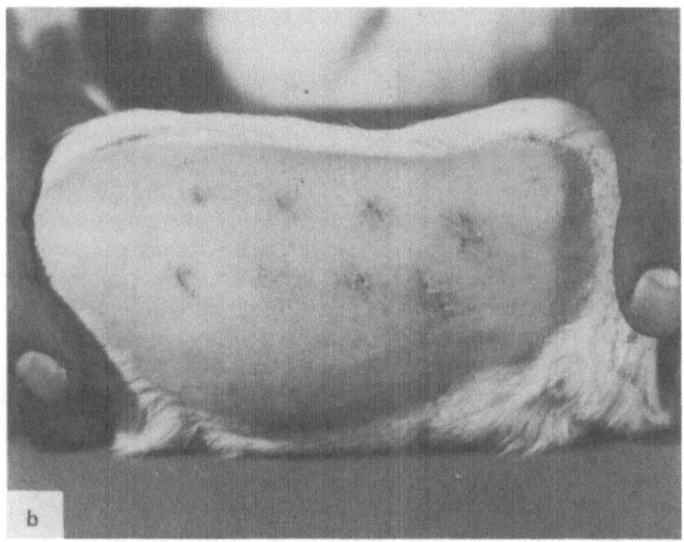

b

sary to try it in much higher concentration in a vehicle which allowed satisfactory penetration before concluding that it has no place in the treatment of genital herpes.

Herpetic keratitis

The follicular conjunctivitis of the primary infection occurs only rarely in recurrent ocular herpes, where the infection gives rise to a dendritic ulcer as the result of progressive destruction of the cells of the corneal epithelium. The virus may also multiply in the cells of the stroma and give rise to disciform keratitis. The course is usually self-limiting, although perforation of the cornea may occur. Corneal opacities may be present after healing and increase in extent after each recurrence, leading ultimately to serious impairment of vision.

The condition frequently recurs. McGill (1975) states that after the first attack of herpetic keratitis 50% of patients will develop a further attack over an 8-year period. In patients with a history of more than one attack the recurrence rate is 80%. Recurrences are much commoner if the preceding attack has been mistakenly treated with steroids.

The standard chemotherapeutic measure in the treatment of herpetic keratitis has for long been the use of idoxuridine in the form of 0·1% eye drops and 0·5% ointment. The treatment was introduced by Kaufman *et al.* (1962 a) and its efficacy was confirmed in a number of subsequent investigations. The most extensive was that of Maxwell (1963), who supplied 0·1% idoxuridine solution to 325 ophthalmologists at various centres in the USA, and analysed the case reports of 465 patients with a first attack of herpetic keratitis who received treatment. Cure was obtained in 152 of 198 patients without stromal involvement, and improvement of the condition was seen in 35 others. Among 168 patients who had a recurrent attack without stromal involvement there were 131 cures and 25 showed improvement. The time to healing in these two groups ranged from 2 to 32 days, with mean values of 5·5 and 7·5 days respectively. In 99 cases with stromal involvement a cure was obtained in 44 and improvement in 45. The mean time to healing was 11·7 days, with a range of 3 to 49 days.

The evaluation of the first trials of idoxuridine in ocular herpes was based upon clinical impression of the response to treatment: Davidson and Evans (1964) investigated its effectiveness in a double-blind trial in which three groups of 25 patients were treated with 0·1% idoxuridine, 1% gamma globulin and iodization respectively. A result assessed as good or excellent was obtained in 14 patients treated with idoxuridine (56%), 13 treated with gamma globulin (52%) and 20 treated by iodization (80%). Idoxuridine was thus less effective than iodization, and the effect observed was not so great as that obtained in the earlier trials. The authors concluded that initial treatment with idoxuridine was justifiable, but recommended that iodization should be carried out if healing did not occur within 7 days. A more favourable result was obtained by Jepson (1964). Improvement was obtained in 11 of 12 patients treated with idoxuridine, as against seven of 12 treated with placebo.

Idoxuridine has become established as an effective treatment for herpetic keratitis, but a number of disadvantages have become apparent (McGill *et al.*, 1974 b). Treatment with idoxuridine does not reduce the recurrence rate. Also, a number of cases fail to respond to treatment, especially if an amoeboid ulcer is present, and this is not usually due to the development of resistance in the infecting virus. Toxic side effects, as described in Chapter 3, may appear and necessitate termination of the treatment. In such cases the use of trifluorothymidine or vidarabine is indicated. Cytarabine is not used in the treatment of herpetic keratitis on account of the toxic effects which it produces in the cornea (Kaufman *et al.*, 1964).

Wellings *et al.* (1972) compared trifluorothymidine and idoxuridine in a double-blind trial carried out on patients with dendritic or amoeboid ulceration of the cornea without associated stromal disease. Treatment consisted of eye drops of 0·1% idoxuridine or 1·0% trifluorothymidine instilled five times a day for 2 weeks. Healing occurred in 37 of 40 (92·5%) patients treated with trifluorothymidine and 23 of 38 (60·5%) treated with idoxuridine. The mean time to healing in the successfully treated cases was 6·3 days and 8·2 days in the two groups respectively. The authors concluded that trifluorothymidine was superior to idoxuridine and possessed the additional advantage of having little or no toxicity.

The use of trifluorothymidine in cases resistant to idoxuridine was studied by McGill *et al.* (1974 a) in a group of 24 patients which included 12 with dendritic ulcers, five with dendritic ulcers also treated with steroids, and seven with amoeboid ulcers treated with steroids. Satisfactory healing was obtained in all cases, with mean times to healing of 5·6 days, 8·6 days and 8·7 days in the three groups respectively. Trifluorothymidine also gave satisfactory results in eight patients who were allergic to idoxuridine. The authors also investigated the frequency of recurrent attacks in the patients treated with idoxuridine and trifluorothymidine in the earlier trial of Wellings *et al.* (1972). Recurrent ulcers occurred in 17 of 21 patients treated with idoxuridine and eight of 14 treated with trifluorothymidine. The difference in frequency was not statistically significant. When the investigation was restricted to patients who were treated for their first ulcer on entering the trial, eight of 12 treated with idoxuridine had recurrences and two of nine treated with trifluorothymidine. The authors consider that the risk of recurrence may be less with trifluorothymidine treatment, although it would be necessary to study a larger number of cases to establish this effect beyond doubt.

Vidarabine has also been used in the treatment of herpetic keratitis. Pavan-Langston and Dohlman (1972) compared its activity with that of idoxuridine in a double-blind trial carried out on 27 patients. Treatment consisted of 3·3% vidarabine ointment or 0·5% idoxuridine ointment given 3–5 times daily. Both treatments were equally effective, and the authors considered that vidarabine is a satisfactory alternative to idoxuridine. Vidarabine has only recently been brought into clinical use, and the outcome of further trials must be awaited before its position in therapy can be considered established.

B virus

B virus is a natural pathogen of rhesus monkeys and it may be present in other species as well. The primary infection produces vesicular lesions on the tongue which soon rupture and give rise to yellowish grey plaques of necrotic tissue. Keeble *et al.* (1958) examined 1400 rhesus monkeys used in laboratory work and found lesions in 32 (2·3%). Involvement of the central nervous system frequently occurs, with lesions in the roots of the facial and trigeminal nerves and solitary tract of the medulla. An examination of 100 monkeys showed that antibody was present in 17.

In man the virus produces an ascending myelitis which is usually fatal. The first case was reported by Sabin and Wright (1934), and occurred in a laboratory worker who was bitten on the fingers by an apparently normal rhesus monkey. Three days after the incident pain, erythema and swelling developed at the site, followed by the formation of vesicles. Seven days later abdominal cramps, hyperaesthesia of the legs and retention of urine appeared, followed next day by flaccid paralysis of the legs. The signs of ascending myelitis progressed and death occurred from respiratory failure on the 16th day of illness. The virus was isolated from the brain, spinal cord and spleen.

Infection may also occur by contact with monkey cages and infected tissue cultures. The disease presents a serious risk to all who work with monkeys. A number of cases have been reported since the original observation of Sabin and Wright and nearly all have been fatal. The pre-existence of herpes neutralizing antibody does not confer protection against infection.

The growth of the virus in tissue culture is inhibited by idoxuridine (Miller, 1967). Formation of plaques by the virus in monkey kidney and HEp-2 cell monolayers is completely inhibited by a concentration of 78 µg/ml (220 µM), and B virus is thus considerably less sensitive to idoxuridine than herpes virus.

Juel-Jensen and MacCallum (1972) have proposed a course of treatment which should be carried out in any cases which may arise in future. The entry wound should be thoroughly cleaned and treated with 40% idoxuridine in dimethylsulphoxide applied on lint. A mixture consisting of 100 mg cytarabine in 10 ml of the idoxuridine solution is also recommended. The application should be renewed daily for 4 days. If B virus encephalitis is suspected systemic treatment with cytarabine or vidarabine must be begun without waiting for the results of virological diagnosis.

GENERAL READING

Illis, L. S. and Gostling, J. V. T. (1972). *Herpes simplex Encephalitis.* (Bristol: Scientechnica (Publishers) Ltd.)

Juel-Jensen, B. E. and MacCallum, F.O. (1972). *Herpes simplex Varicella and Zoster.* (London: William Heinemann Medical Books, Ltd.)

Chapter 6

Chemotherapy of herpesvirus infections—II

Varicella-zoster and cytomegalovirus

HERPESVIRUS INFECTIONS OF SUBGROUP B

Varicella-zoster

Varicella and zoster are different manifestations of infection with the same agent, which is generally referred to as varicella-zoster virus. The identity was first shown by Weller and Coons (1954) in immunofluorescence studies carried out with antisera from patients who had suffered from the two diseases. Cultures of human embryo skin-muscle cells infected with strains of virus isolated from the lesions of varicella and zoster stained equally well with both antisera and it was thus not possible to detect any antigenic difference between them.

Primary infection

The primary infection usually occurs in childhood and gives rise to the familiar condition of varicella. A vesicular eruption appears without a prodromal phase. The distribution is centripetal and the lesions characteristically appear in crops, so that different stages in their evolution are present at the same time. In adults the disease is more severe. There is usually a prodromal period of 2–3 days, with headache, fever, backache and other pains. The rash is usually much more profuse than in children.

The disease is not serious enough to warrant treatment with the agents currently available, but in patients who have congenital defects of immunity or are being treated with immunosuppressants the disease is much more severe and may be life-threatening. In such cases specific treatment is indicated.

Hall *et al.* (1969) investigated the effect of cytarabine in two children with congenital immunological deficiencies and five with malignant conditions. It was given to one patient in a dose of 100 mg/m² as an intravenous infusion over a period of 24 hours. Crusting of the lesions began after 36 hours and treatment was stopped after a total dose of 75 mg had been given. Recovery was complete 6 days after beginning treatment. A second patient was given cytarabine in doses of 3 mg/kg on the 5th and 1½ mg/kg on the 6th day of illness, and crusting of the lesions began on the following day. However, new vesicles appeared 1 week later. A second course of treatment was ineffective and death occurred 14 days after the onset of the second attack. The dose given in the remaining five patients is not stated. One died 36 hours after the beginning of treatment but there was a prompt response in the others.

Chow *et al.* (1970) used cytarabine in the treatment of varicella occurring in a child with leukaemia. It was given as a continuous intravenous infusion in a dose of 80 mg/m² for 2½ days, followed by 60 mg/m² for 1 day and 100 mg/m² for 4 days. A response to treatment was observed within 12 hours, and healing was complete within 7 days. A similar result was reported by Prager *et al.* (1971). A girl aged 6 years developed varicella while undergoing treatment of leukaemia with cytotoxic and other agents. Cytarabine was given in the form of rapid intravenous infusions of 100 mg/m² each day for 7 days. The condition improved rapidly and the lesions had mostly disappeared after 3 days of treatment.

An investigation of the effects of cytarabine in herpesvirus infections by Hryniuk *et al.* (1972) includes a case of fulminant varicella occurring in a child aged 2½ years who was suffering from acute myeloblastic anaemia treated with vincristine, prednisone and X-irradiation. Cytarabine was given by intravenous infusion in a daily dose of 80 mg/m² for 2½ days, followed by 60 mg/m² for 1 day and 100 mg/m² for 4 days. Improvement set in within 12 hours of beginning treatment and healing was complete after 1 week. The authors also reviewed other cases reported in the literature and concluded that treatment with cytarabine should be given in life-threatening infection, but that the dose should be carefully chosen so as to exert an antiviral effect without giving rise to toxicity or immunosuppression.

Vidarabine has recently been evaluated in the treatment of varicella and related infections (Johnson *et al.*, 1975). It was given by intravenous infusion over a period of 6 hours in doses of 5–20 mg/kg for 5 to 7 days. A good response to treatment was obtained in six of nine patients with varicella, including three with chronic obstructive disease of the lungs, one with chronic lymphocytic leukaemia and two with no apparent underlying disease. The remaining patients died, two with progressive varicellar pneumonia and one with Reye's syndrome. The authors considered that vidarabine was probably effective but that further investigation would be necessary before its place in the treatment of varicella-zoster can be assessed.

Varicellar pneumonia

Pneumonia due to infection of the lungs with varicella-zoster virus is an occasional complication of varicella, particularly in young adults, in whom it may be severe and sometimes fatal. Among the patients with varicella treated with cytarabine in the series reported by Hall *et al.* (1969) there were five with varicellar pneumonia. The patients were children with acute lymphocytic leukaemia or other malignant conditions. Cytarabine was given to one patient as a 24-hour infusion of 100 mg/m^2 to a total dose of 75 mg. Improvement set in rapidly and the patient recovered. A second child was treated with 3 mg/kg on the 5th day and 1$\frac{1}{2}$ mg/kg on the 6th day of illness. There was a prompt response to treatment but there was a recurrence of infection a week later which was unaffected by a further course of treatment, and death occurred 29 days after the onset of the initial illness.

The cases of varicella treated with vidarabine by Johnston *et al.* (1975) included seven with varicellar pneumonia. Five recovered, but death occurred in the remaining two. The dose given ranged from 5 mg/kg to 20 mg/kg per day. Further studies will be necessary before the effectiveness of vidarabine in the treatment of varicellar pneumonia can be established.

Varicella encephalitis

Encephalitis is an infrequent complication of varicella. No reports on specific treatment have appeared so far, but Juel-Jensen and MacCallum (1972) consider that the systemic administration of a chemotherapeutic agent is indicated.

Latency and recurrence

Information on the behaviour of varicella-zoster virus during its latent phase is less comprehensive than in the case of herpes virus on account of the absence of an experimental animal model. Cheatham *et al.* (1956) investigated the distribution of the virus in autopsy material from a child with a metastasizing neuroblastoma who developed generalized varicella while undergoing treatment with methotrexate. Most internal organs showed signs of infection, and the characteristic inclusion bodies of varicella-zoster were found in the dorsal root ganglia of segments T6 and T9–T11. These findings show that the virus can become established in the regional sensory ganglia during the course of the primary infection.

A similar localization preceding recurrence was observed by Esiri and

Tomlinson (1972) in a patient with myeloma who died 4 days after developing zoster in the distribution of the ophthalmic division of the right trigeminal nerve. Examination in the electron microscope of materials removed at autopsy revealed the presence of virus particles in the bundles of the frontal nerve. They were present in the cytoplasm of the perineurial cells and in both the cytoplasm and the nuclei of the Schwann cells. No virus particles could be found in the axons. In the trigeminal ganglion virus particles were present in the cytoplasm and nuclei of the neurones and satellite cells. The virus could not be isolated in tissue culture but was identified as varicella-zoster by immunofluorescence. The authors concluded that varicella-zoster virus spreads down nerve trunks by progressive infection of the perineurial cells and not by transport down the axons.

In the study of 61 immunosuppressed renal transplant recipients carried out by Lopez *et al.* (1974) varicella-zoster virus was found in seven, but there was no information on the site of latency.

Recurrent infections with varicella-zoster virus take the form of localized and generalized zoster, zoster keratitis and zoster encephalitis.

Localized zoster

The eruption of localized zoster occurs in the dermatomes innervated by the dorsal root ganglia and nerve trunks in which the virus is presumably lying latent. The nature of the provoking stimulus was investigated by Juel-Jensen (1970) in a group of 100 patients. An associated event could be implicated in 65. This was most frequently physical trauma (38 patients), and there was also a notable association with frontal sinusitis (nine patients). Pain usually develops in the affected area and is followed by a vesicular eruption which resembles that of varicella in its appearance and evolution. The condition is self-limiting, but may be followed by long-lasting neuralgia which may be of great severity.

The good results obtained by the use of topical applications of idoxuridine in cutaneous herpes led Juel-Jensen *et al.* (1970) to investigate its effect in localized zoster. In a group of seven patients a 5% solution in dimethylsulphoxide was applied to the lesions four times a day with a paint brush. In two control groups seven patients were similarly treated with dimethylsulphoxide and ten with saline. The median duration of pain was 3, 13·5 and 30 days in the three groups respectively, but there was no significant difference in the time to healing. In a continuation of the trial a 40% solution in dimethylsulphoxide was applied continuously on lint which was rewetted daily for 4 days. Two control groups were set up as before, and the numbers of patients were nine, seven and four in the three groups respectively. The median duration of pain was 3 days in the group treated with idoxuridine, compared with 65 and 14·5 days in the other groups. There was no significant difference in the time to healing, but the reduction in the duration of pain was so marked that the authors concluded that continuous treatment with 40% idoxuridine had a definite place in the treatment of zoster.

In a double-blind trial carried out by Dawber (1974) good results were obtained with lower concentrations. The mean duration of pain was 9·8 days in a group of 19 patients treated with a 5% solution 4-hourly, and 9·2 days in 19 patients treated similarly with a 25% solution, compared with 19·1 days in 20 patients treated with dimethylsulphoxide alone. The treatment also produced some reduction in the duration of the vesicular phase.

Zoster has also been treated with cytarabine, but the effectiveness of treatment has not been established beyond doubt. It was first used by Chow et al. (1970). Four patients were treated with cytarabine given by continuous infusions of 20 mg/m^2 for periods up to 3 days. A favourable response occurred within 12 hours; healing was complete in 4–5 days and the pain subsided in 0·5–1·5 days, except in one patient who had pain for 42 days. Similar results were reported by Hryniuk et al. (1972) in five patients with localized zoster. Cytarabine was given by continuous intravenous infusion in daily doses of 10 or 20 mg/m^2 for 1½–3 days. The infection was arrested in all cases within 36 hours and pain subsided within the same period, except for two patients in whom it persisted for 42 and 63 days respectively. The treatment was well tolerated and myelotoxic effects were limited to a fall in reticulocyte count and megaloblastic changes in the bone marrow.

Other reports have been unfavourable, since zoster may occur during treatment with cytarabine, and double-blind controlled studies have given negative results. Mann (1971) observed an attack of zoster in a 13-year-old boy with acute lymphoblastic leukaemia who was being treated with cytarabine, and concluded that the treatment was ineffective. However, the patient was in remission and receiving a weekly maintenance dose of 80 mg/m^2. On account of the short plasma half-life the compound would not be present in an adequate antiviral concentration and infection would therefore not be prevented. Davis et al. (1973) carried out a double-blind trial in eight patients with Hodgkin's disease and other malignant conditions. Five patients received cytarabine in a dose of 150 mg/m^2 given over 36 hours by continuous intravenous infusion, and three were given 5% glucose saline as a placebo. One patient treated with cytarabine developed disseminated zoster, and there was no difference between the two groups in the length of the acute phase and the time to healing.

A similar trial was carried out by Schimpff et al. (1974). Cytarabine was given to 17 patients in a daily dose of 30 mg/m^2 by continuous infusion over a period of 3 days, and eight patients received a placebo. There was no evidence of clinical effect. In four patients treated with cytarabine the lesions progressed in the affected dermatomes and new vesicles appeared;

dissemination occurred in one treated patient and there was no reduction in the duration of pain.

A more extensive study carried out by Betts *et al.* (1975) also gave negative results. The same group of workers (Zaky *et al.*, 1975) had shown earlier that blood levels exceeding 0·125 μg/ml were present for a period of 75 minutes when cytarabine was given to 12 patients in a dose of 50 mg/m^2 by the subcutaneous route. A study of 26 strains of varicella-zoster virus by the plaque reduction method showed that the 50% inhibiting concentration ranged from 0·00156 to 0·25 μg/ml, and subcutaneous injection thus afforded an effective means of attaining an antiviral concentration. In a double-blind study in which this method was used 30 patients with localized zoster were given a subcutaneous injection of 50 mg/m^2 each day for 4 days and 30 closely-matched patients were given a placebo. No differences in clinical course could be detected between the two groups. Determinations of blood levels carried out on 12 patients showed that cytarabine could still be detected in a concentration of 0·086 μg/ml 90 minutes after injection. The authors concluded that cytarabine was ineffective in the treatment of localized zoster, but it is evident that adequate blood levels would be present for only a fraction of the replication cycle of the virus, so that no therapeutic effect could be expected from treatment given only once a day.

It seems likely that cytarabine would be effective if applied topically as a solution in dimethylsulphoxide, but there are no reports of its having been used in this way. There are also no reports on the use of trifluorothymidine.

Johnson *et al.* (1975) made a limited study of the effect of vidarabine in the treatment of localized zoster in 14 patients who were mostly suffering from malignant conditions and under treatment with immunosuppressive agents. It was given intravenously in a daily dose of 10–20 mg/kg over a 6-hour period for 5–7 days. New lesions ceased to appear by the 4th day of treatment. The authors concluded that further trials would be necessary to establish whether vidarabine had any place in treatment.

Varicella-zoster keratitis

The lesions of varicella may rarely affect the cornea. The course is usually benign but there may be some residual opacity. Cairns (1964) reported a case treated with 0·5% idoxuridine ointment given every 2 hours for 5 days. The lesion regressed during the second day of treatment.

The eye is affected in about 50% of cases of zoster occurring over the distribution of the ophthalmic division of the trigeminal nerve. The lesions take the form of vesicles in the corneal epithelium and disciform involvement of the stroma.

Pavan-Langston and McCulley (1973) have recently observed three cases of dendritic ulceration of the cornea resembling that seen in herpetic keratitis. The condition responded to treatment with idoxuridine ointment. Kaufman *et al.* (1963) observed three patients who developed dendritic ulcers during treatment of ophthalmic zoster with steroids. The lesions improved rapidly when treated with 0·1% idoxuridine eye drops.

Disseminated zoster

In patients suffering from malignant conditions who are treated with steroids and immunosuppressive agents the recurrence of varicella-zoster infection may present as a generalized infection, sometimes with involvement of the lungs and encephalitis. Topical treatment with idoxuridine solution will not reach all sites of infection, and systemic treatment is therefore indicated. Treatment with cytarabine has often been advocated, but there is some doubt as to its effectiveness. The first instance of its use is reported by McKelvey and Kwaan (1969), who observed the development of generalized zoster during an episode of lobar pneumonia occurring in a man with a lymphoproliferative disorder of uncertain nature who was undergoing treatment with melphalan and prednisone. The eruption was very extensive and the condition began to deteriorate. On the 3rd day cytarabine was given as a continuous intravenous infusion of 100 mg/m^2 (180 mg) in each 24-hour period over a total period of 120 hours. An improvement in the general condition set in within 24 hours and by the 3rd day the disease process was arrested.

The study by Chow *et al.* (1970) discussed earlier included one case of disseminated zoster which responded to cytarabine given according to a similar schedule in a daily dose of 40 mg/m^2 for 6½ days. The infection was arrested on the first day of treatment. Hryniuk *et al.* (1972) used cytarabine in the treatment of two patients with disseminated zoster, given as a continuous infusion of 10 and 40 mg/m^2 for 1½ and 6½ days respectively. In the patient on the higher dose the infection was arrested and the pain subsided on the first day of treatment. In the other patient these responses were obtained after 1½ and 5 days respectively.

Other reports on the treatment of disseminated zoster with cytarabine have been unfavourable. The occurrence of dissemination in patients with localized zoster who received 150 mg/m^2 per day has already been mentioned (Davis *et al.*, 1973). Stevens *et al.* (1973) carried out a double-blind controlled trial of cytarabine in 39 patients. Nineteen patients received a daily dose of 100 mg/m^2 for up to 72 hours, and 20 received a placebo. Both groups were matched for nature of the underlying disease, and treatment was begun within 48 hours of the onset of dissemination. The mean duration of dissemination in treated and placebo groups was 3 and 4 days respectively, and the mean length of stay in hospital was 9·4 and 5·6 days. It was concluded that cytarabine had no therapeutic effect and actually made the condition worse. In both these studies cytarabine was given in doses which were high enough to cause immunosuppression, and the results cannot be regarded as a failure to confirm the good results obtained with much lower doses by Hryniuk *et al.* (1972).

Juel-Jensen and MacCallum (1972) considered that cytarabine was the drug of choice in the treatment of disseminated zoster. On the basis of their experience of 30 cases, including a number with generalized herpes, they recommended an initial dose of 5 mg/kg, followed by 3 mg/kg daily for 3 days. If the marrow is markedly depressed the dose should be reduced to 2 or 2·5 mg/kg per day.

Waltuch and Sachs (1968) described a case which was successfully treated with idoxuridine. The patient was a man aged 34 with widespread Hodgkin's disease of 11 years duration. After exposure to a patient with varicella a zoster eruption appeared on the left side in the distribution of C1 and C2. Six days later dissemination occurred, with lesions on the face, trunk and extremities. On the 9th day after onset idoxuridine was given in a dose of 80 mg/kg, repeated on the 2 following days. The treatment was then suspended for 1 day and a final dose of 40 mg/kg was then given. The total dose was thus 25·2 g over a period of 5 days. By the second day of treatment new lesions had ceased to appear. Improvement set in and healing was complete by 11 days after the beginning of treatment. In spite of this favourable result idoxuridine does not seem to have come into general use in the treatment of disseminated zoster.

Vidarabine was used by Luby *et al.* (1975) in the treatment of 19 patients with disseminated zoster. The dose was 5–15 mg/kg infused intravenously over a period of 6 hours on 5 successive days. New lesions ceased to appear after 4 days of treatment. Three patients died, and the authors considered that a double-blind trial would be necessary to establish whether vidarabine was effective.

Zoster pneumonia

Pneumonia due to infection with varicella-zoster virus is an occasional complication of generalized zoster. The treatment is the same as that of varicellar pneumonia. Baron and Wechsler (1975) described a case which occurred in a man aged 56 with mycosis fungoides in remission who was on maintenance therapy with triamcinolone. A zoster eruption appeared on the right shoulder and subsequently became disseminated to involve the whole body. Pneumonia developed and the patient became comatose. Cytarabine was given intravenously in an initial rapid dose of 30 mg, followed by 15 mg in 100 ml of 5% glucose saline by continuous drip every 12 hours for 5 days, the total dose being 180 mg. The patient recovered rapidly. The authors ascribed the good therapeutic result to the use of low doses of cytarabine which made it possible to avoid toxic effects upon the bone marrow.

Zoster encephalitis

Encephalitis is an occasional complication of disseminated zoster. Juel-Jensen and MacCallum (1972) reported two cases in which treatment with cytarabine seemed to be effective. The dose was 2 mg/kg per day for 5 days in one case and 3 mg/kg daily for 5 days in the other. Both patients made a rapid recovery from the virus infection, but the second patient died a few days later from bacterial bronchopneumonia.

There have been no reports on the use of idoxuridine or vidarabine. Johnson *et al.* (1975) observed the onset of encephalitis in a patient with zoster on the 3rd day of treatment with vidarabine, and it is therefore unlikely that vidarabine would be effective in the treatment of this condition.

Cytomegalovirus

Cells infected with cytomegalovirus are enlarged and contain intranuclear and intracytoplasmic inclusions. This characteristic appearance has been known for years, long before it was found to be due to a virus infection. Farber and Wolbach (1932) examined salivary glands taken at autopsies on 183 children and found the characteristic inclusions in 22 (12%). They also summarized earlier reports in the literature which covered 25 cases. The inclusions were found in the kidneys in 11, the parotid glands in ten, and in the lungs and liver in eight cases.

Infection with cytomegalovirus is widespread among the general population. Rowe *et al.* (1956) found that specific antibody was present in 12 of 17 (71%) infants at birth, representing maternal antibody which had crossed the placenta. The incidence fell to three of 21 (14%) between 6 months and 2 years, and then rose steadily to attain a value of 45/52 (81%) in persons aged 35 and over. Similar results were obtained by Stern and Elek (1965) who studied a population of normal adults and children suffering from diseases unrelated to cytomegalovirus infection, who were resident in the Greater London area. The incidence reached a maximum of 54% (62 of 114) between the ages of 25 and 35 years. Deibel *et al.* (1974) examined the sera of 4869 persons of all ages who were resident in New York State. The frequency of antibody began to rise from the age of 3 years onward, and attained a maximum of 92% in persons over 50. It is evident that the incidence of infection varies in different population groups, and also that primary infection with cytomegalovirus can be delayed until late adult life.

The distinction between primary and recurrent infections is much less marked with cytomegalovirus than with herpes and varicella-zoster. The types of infection which it causes are collectively referred to as cytomegalic inclusion disease.

Primary infection

Primary infection with cytomegalovirus is often asymptomatic. Levinsohn *et al.* (1969) examined 100 apparently normal infants on five occasions during the first year of life for the presence of cytomegalovirus in throat swabs and urine. The virus was isolated from 15, and once it had appeared it was excreted continuously. In spite of the continued presence of infection the general health and development remained unaffected.

Numazaki *et al.* (1970) carried out a similar study on 257 healthy Japanese infants living in their own homes. Mouth swabs were taken at intervals up to the age of 2 years and investigated for the presence of cytomegalovirus. It was isolated from 52 (20%) of 257 over the whole period. It could not be isolated from 30 infants examined at birth, but mostly appeared between the ages of 5 and 9 months. Complement-fixing antibody appeared in 60% between the ages of 6 and 12 months, a finding which showed that an asymptomatic primary infection had taken place.

A primary or recurrent infection occurring in pregnant women may lead to generalized infection of the fetus, with extensive involvement of the central nervous system and internal organs. The condition is known as congenital cytomegalic inclusion disease. A typical description of the condition has been given by McCracken *et al.* (1969), who observed 18 such cases for periods up to 9 years. The liver was involved in all cases, with hepatomegaly and abnormal liver function. In some patients the condition reversed or showed some improvement. The central nervous system was involved in nine patients, with microcephaly, cerebral calcification and mental retardation. In five patients there was congenital heart disease and abnormalities of the derivatives of the first arch, including arched or cleft palate, micrognathia and left facial weakness, which indicated that the virus had a teratogenic effect.

Primary cytomegalovirus infection can also be acquired from blood, particularly when large amounts are given, as in exchange transfusion and the use of extracorporeal circulation in open heart surgery. In such circumstances it may give rise to a condition resembling infectious mononucleosis. Three such cases were described by Kääriänen *et al.* (1966). The illness usually comes on 4–10 weeks after the administration of blood, and can be distinguished from infectious mononucleosis by the absence of heterophil agglutinins. Carter (1968) reported a case which took a form resembling infective hepatitis; the correct diagnosis was established by the finding of complement-fixing antibody to cytomegalovirus in the serum in a titre of 4096. Evidence of the presence of the virus in the blood of apparently healthy persons has been provided by Diosi *et al.* (1969), who isolated it from cultures of the leukocytes of two of 32 blood donors who gave no history of recent illness.

The results of treatment of the various aspects of primary cytomegalovirus infections have been far from satisfactory. No treatment is indicated for healthy persons who are found to be excreting the virus, and in congenital cytomegalic inclusion disease the damage to the organs which has already occurred is largely irreversible.

Conchie *et al.* (1968) used idoxuridine in the treatment of a case of congenital cytomegalic inclusion disease. The patient showed slight microcephaly, cerebral calcification and increased muscle tone. The virus was present in the urine. Treatment was begun in the 14th week of life, with 200 mg/kg given by intravenous infusion, followed by 100 mg/kg on each of the 4 following days. The titre of virus in the urine fell by 3 log units. At 10 months the patient was feeding well, but spasticity was still present and there was little progress in motor development. Barton and Tobin (1970) treated three patients, whose symptoms included microcephaly, splenomegaly, jaundice and purpura. The first patient received idoxuridine on the 14th day of life in an initial dose of 200 mg/kg given intravenously over a period of 2 hours, followed by 100 mg/kg on each of the 4 following days. The second patient received similar treatment, except that the lower dose was carried on for 6 days. Both patients showed slight general improvement during treatment. The titre of virus in the first patient fell by 2 log units, but there was no change in the second. The third patient was treated at the age of $2\frac{1}{2}$ years. Idoxuridine was given in a total dose of 595 mg/kg in equal doses over a period of 7 days as intravenous injections given over a period of 2 hours. Virus was eliminated from the urine, but the titre before treatment was not particularly high.

Treatment with cytarabine has been no more satisfactory. Plotkin and Stetler (1969) used it in the treatment of three patients whose symptoms at birth included jaundice, hepatosplenomegaly and chorioretinitis. The first patient was treated in the 7th week of life with doses of 15 mg/kg given by intravenous injection daily for 5 days. There was a slight improvement in the general condition. The titre of virus in the urine before treatment was 5 log units. It disappeared during treatment and was still absent 1 year later. The second patient was treated in the 6th week of life with a daily dose of 10 mg/kg for 3 days. The general condition improved and the patient began to gain weight. The virus disappeared from the urine but reappeared 3 months later. At 7 months of age the patient was microcephalic but otherwise healthy and thriving. The third patient had hepatosplenomegaly at birth, and at 4 months of age there was microcephaly and chorioretinitis. The virus was still being excreted in the urine at the age of 22 months. Cytarabine was then given in doses of 9 mg/kg daily for 5 days. The excretion of virus was not affected, but 10 weeks later the patient had gained 1200 g in weight and the circumference of the head had increased by 1·5 cm.

Cytarabine was also used by McCracken and Luby (1972), who treated four patients with hepatosplenomegaly. The first patient was treated at the age of 4 months and received cytarabine in a total dose of 58 mg in three courses, the first being 2 mg/kg daily for 5 days, followed by 8 mg/kg daily for two periods of 3 days. The titre of virus in the urine fell by 4·5 log units, but returned to the original value at the end of the treatment. Death occurred at the age of 7 months with disseminated cytomegalovirus infection. The second patient was treated at the age of 3 months with 8 mg/kg daily for 5 days. There was no change in the general condition, but the liver and spleen were reduced in size 30 days later. The titre of virus in the urine fell by 2 log units, but then returned to the initial value. At 8 months of age the liver and spleen showed a further decrease in size, there was a delay in motor development and virus was still present in the urine. The third patient was treated with 8 mg/kg daily for periods of 5 and 3 days separated by an interval of 4 days. There was a fall of 2 log units in the titre of virus in the urine. At 6 months the liver and spleen were still enlarged, and there was also microcephaly and retardation of development. Virus was present in the urine in high titre. The fourth patient received a single dose of 25 mg/ kg given intravenously over a period of 2 hours and repeated 2 weeks later. There was a moderate degree of clinical improvement and the titre of virus in the urine fell by 3 log units during treatment.

The largest group of cases yet reported is that of Baublis et al. (1975), who treated 17 cases of congenital cytomegalovirus infection with vidarabine. The dose ranged from 1 to 20 mg/kg per day given over a period of 5–21 days. The degree of suppression of excretion of virus in the urine ranged from nil to 4 log units. Subjective clinical improvement was obtained in 11 patients; four patients died, two from conditions not directly related to cytomegalovirus infection. The condition at follow-up ranged from a normal developmental state to severe retardation.

Cytomegalic inclusion disease has also been treated with floxuridine, the fluorine analogue of idoxuridine (Cangir et al., 1967; Feigin et al., 1971), but this compound has not come into general use.

The above cases have been described in some detail in order to illustrate the lack of consistency in the treatment régimes and in the results obtained. It can be inferred that arrest of the infection could best be attained by instituting treatment as soon as the diagnosis is established, and pursuing it over a much longer period subject to adequate monitoring of toxic side effects.

Latency and recurrence

Little is known of the behaviour of cytomegalovirus during the latent phase of the infection. The detection of the virus in the leukocytes of healthy blood donors has already been mentioned (Diosi et al., 1969), and also the frequent observation of characteristic enlarged cells with inclusions at autopsy (Farber and Wolbach, 1932). Feldman (1969) found that six of 185 (3%) apparently healthy pregnant women were excreting the virus.

The virus therefore appears able to produce a persistent low level of infection which is not eradicated by immune mechanisms. The virus may also persist by integration with the host cell genome, since Joncas *et al.* (1975) found that cytomegalovirus DNA could be detected in the DNA of a lymphoblastoid cell line established from an infant with congenital cytomegalovirus infection. Little is known of the mechanisms which lead to recurrences of the infection.

Recurrent infection

Reactivation of cytomegalovirus infection may occur in patients undergoing immunosuppressive treatment, and in pregnancy. Hill *et al.* (1964) found pulmonary infection with cytomegalovirus in 20 of 39 patients who died after receiving organ transplants and immunosuppressive therapy. In similar circumstances Rifkind *et al.* (1964) observed the development of pneumonitis 42–102 days after operation in six patients who had received kidney transplants; one case was fatal. It was uncertain whether these cases were due to primary infection or reactivation, but in a number of reports the infection was observed to develop in patients who had initial antibody. Craigland (1969) observed a rise in titre of complement-fixing antibody to cytomegalovirus indicative of reactivation in 29 of 45 patients who possessed antibody before undergoing renal transplant operations. Nagington (1971) found that 42 of 50 (84%) patients already had antibody on admission to a renal transplant unit. Evidence of active infection was observed in 43 after admission. He concluded that primary infection was rare in transplant patients, and that cytomegalovirus infections are usually reactivations brought about by immunosuppressive therapy.

Reactivation may occur after blood transfusion or the use of extracorporeal circulation. Foster and Jack (1969) observed the development of post-transfusion mononucleosis with fever and atypical mononuclear cells in six patients who had had open heart surgery. Antibody to cytomegalovirus was present before operation in all cases, and the virus was isolated from the peripheral blood leukocytes of two patients.

Duvall *et al.* (1966) isolated cytomegalovirus from 11 of 32 adult patients with neoplastic disease. Antibody was present when excretion of virus was first detected, but it was not certain whether the infection was a reactivation or not. It was usually associated with hypogammaglobulinaemia and the use of corticosteroids.

Pregnancy frequently acts as a stimulus for the reactivation of cytomegalovirus infection. This is nearly always asymptomatic, but it is the usual cause of congenital cytomegalic inclusion disease in the infant. Hildebrandt *et al.* (1967) isolated cytomegalovirus from the urine of seven of 210 pregnant women at delivery. A higher frequency of infection was found by Montgomery *et al.* (1972), who recovered the virus from the cervix of 14 (8%) of 176 pregnant women. The virus was detected most frequently in the third trimester. Reynolds *et al.* (1973) obtained similar results, and recovered the virus from the cervix of 35 (13·4%) of 261 pregnant women in the third trimester.

Most recurrences of cytomegalovirus infection are asymptomatic or too mild to require treatment, but they are occasionally severe or fatal. There have been no reports of the use of idoxuridine or cytarabine in treatment. Two of the cases studied by Duvall *et al.* (1966) occurred in patients who were already under treatment with cytarabine, but the treatment régime used in malignant conditions would probably not be suitable for treating a virus infection. The series of cases treated with vidarabine reported by Baublis *et al.* (1975) included six adult patients suffering from cytomegalovirus mononucleosis and hepatitis. Vidarabine was given intravenously in a dose of 10 mg/kg daily for 10 or 14 days. Two patients showed subjective clinical improvement but the condition of the others was unchanged by treatment.

GENERAL READING

Juel-Jensen, B. E. and MacCallum, F. O. (1972). *Herpes simplex Varicella and Zoster*. (London: William Heinemann Medical Books, Ltd.)

Chapter 7

Chemotherapy of poxvirus infections

The poxviruses are classified in six subgroups, as shown in Table 7.1, which is modified from one given by Andrewes and Pereira (1972). The viruses which infect man are vaccinia, smallpox in its two variants variola and alastrim, cowpox, monkeypox, orf, paravaccinia, molluscum contagiosum, bovine papular stomatitis, camel pox, Yaba virus and Tana pox. The poxviruses of importance in chemotherapy are vaccinia, variola and alastrim.

The success of the World Health Organization's campaign for the eradication of smallpox has resulted in the virtual disappearance of indications for treating poxvirus infections of man. The disease has been eradicated from all countries except for an inaccessible region in Ethiopia from which it is unlikely to escape. There is therefore no further indication for general vaccination against smallpox except in field workers exposed to the disease, and vaccinia infections requiring specific treatment will eventually cease to exist. The following account of the chemotherapy of the poxvirus infections is therefore presented as a record of part of the history of antiviral chemotherapy, but may nevertheless be of value in the case of a possible recrudescence of smallpox sometime in the future.

Vaccinia

Vaccination against smallpox consists of the production of an infection with vaccinia virus, which is introduced into the skin, usually with a bifurcated needle or by scarification. Vaccinia virus does not exist under natural conditions and its origin is obscure. The strains used for vaccination in different countries differ somewhat in their properties, and they may have been derived independently as a result of recombination between variola and cowpox at various times during the 19th century.

The first infection with vaccinia is referred to as primary vaccination. The immunity which it induces declines with time and must then be supplemented by revaccination.

Table 7.1   Classification of the poxviruses (Andrewes and Pereira, 1972)

Subgroup 1	Subgroup 2	Subgroup 3	Subgroup 4	Subgroup 5	Subgroup 6
*Vaccinia	*Orf	Sheeppox	Bird poxes	Myxoma	*Molluscum contagiosum
*Variola	*Paravaccinia	Goatpox		Fibroma	*Tanapox
*Alastrim	*Bovine papular stomatitis	Lumpy skin disease			*Yaba
*Monkeypox					*Camelpox
*Cowpox					

* Viruses affecting man

Primary vaccination

The lesion of primary vaccination usually pursues an uneventful course, but treatment is indicated in certain circumstances. Primary infection may occur in an undesirable site, such as the nasal cavity or orbit, as a result of autoinoculation. Also, it may sometimes be necessary to carry out primary vaccination in the presence of complications. In such cases treatment with methisazone may be given. Jaroszyńska-Weinberger (1970) investigated the effect of treatment with methisazone on 16 children with ectopic primary lesions of the orbit, tongue, buccal mucosa, nose and mouth. An initial dose of 100–150 mg/kg was followed by 150 mg/kg given daily for 3–7 days. The total dose ranged from 3·2 g to 14·0 g. The mean time to complete healing was 13·3 days, whereas it was 17·8 days in a control group of similarly infected children not treated with methisazone. Jaroszyńska-Weinberger and Mészáros (1966) investigated the effect of methisazone in a group of 26 children with various contraindications in whom primary smallpox vaccination was nevertheless required. It was given in an initial dose of 100 mg/kg given on the 4th day after vaccination, followed by 50–60 mg/kg daily for 3–6 days. There was no untreated control group, since the vaccination of such children is a risky procedure, and the results of treatment were therefore evaluated in comparison with a group of 29 children with similar contraindications who were vaccinated and protected with anti-vaccinial gamma-globulin. The dimensions of the lesion of primary vaccination were markedly reduced in the group treated with methisazone, and the difference from the gamma-globulin group was highly significant ($p = 0·005$). Treatment with methisazone did not affect the development of neutralizing antibody in response to vaccination, since determinations carried out at intervals up to 9 weeks after vaccination in 19 and 22 children from the two groups respectively gave geometric mean titres of 172·1 and 193, a difference which was not significant ($p > 0·5$). Bondarev et al. (1970) investigated the effect of methisazone in 110 children with various contraindications who were given primary smallpox vaccination. A group of 52 similar children was vaccinated and left untreated for control purposes. The methisazone used was manufactured in the Soviet Union, and was given as tablets in a dose of 10 mg/kg twice daily, beginning at various times after vaccination. The reaction to vaccination was mild in 17 of 20 children treated for 4 days after vaccination, and in 36 of 48 treated on days 5–10, compared with 26 mild reactions among the 52 control children. The difference in favour of the treatment was highly significant ($p < 0·02$). There was little difference in the maximum temperature observed, but the duration of the febrile period was reduced in the methisazone group ($p < 0·05$) and the area of the local lesion was smaller ($p < 0·05$). The authors concluded that treatment with methisazone reduced the severity of the local and general reactions to vaccination, and recommended that it should be given between the 5th and 8th days after vaccination. The formation of neutralizing antibody was unaffected. The effect obtained is rather surprising in view of the low doses given.

During the course of a WHO vaccination campaign in the Equatorial Province of Zaire, a region affected by smallpox, primary vaccination was necessarily carried out on children suffering from contraindications such as kwashiorkor, yaws, eczema, pemphigus, sycosis, ichthyosis, psoriasis and scabies. With the object of preventing infective complications of vaccination, Ladniy (1974) treated 279 children with Soviet methisazone in the form of tablets given in a dose of 0·3 or 0·6 g daily for 4 days, beginning 4 days after vaccination. Complications, which included eczema vaccinatum, generalized vaccinia, autoinoculation and encephalopathy, occurred in 17 children (6·1%), whereas in an untreated control group of vaccinated children with similar contraindications complications occurred in 55 children (18·5%). There were two and eight fatal cases in the two groups respectively. There was thus a 3-fold reduction in the incidence of complications in the group treated with methisazone, which was statistically significant. In addition to the above complications two cases of vaccinia gangrenosa occurred in the untreated group. On the basis of these results Ladniy recommended the use of methisazone for prophylactic purposes when primary vaccination must be carried out in the presence of contraindications.

Eczema vaccinatum

If persons suffering from eczema are vaccinated against smallpox the vaccinia infection may not remain confined to the site of inoculation but may spread to involve the eczematous areas and other parts of the body, producing a generalized eruption of vesicles and pustules with fever and severe general reaction. The condition resembles smallpox in many ways and may be fatal in the absence of treatment. It may occur whether the eczema is active or not. It was originally described as Kaposi's varicelliform eruption, but is now known as eczema vaccinatum. It commonly occurs in infants with infantile eczema. Vaccination is usually withheld in such cases, but the patient often becomes infected from a vaccinated parent or sibling.

The specific treatment of eczema vaccinatum with methisazone was first reported by Turner et al. (1962). The patient was an eczematous infant who was accidentally infected with vaccinia at the age of 7 months by contact with a recently vaccinated parent. A generalized eruption developed, with fever, prostration and toxaemia. Treatment with antivaccinial gamma-globulin was ineffective, but the condition rapidly responded to treatment with methisazone given orally in a dose of 250 mg 6-hourly.

The clinical course of eczema vaccinatum is very variable, and assessment of the effect of treatment can only be based on clinical impressions.

Table 7.2 Treatment of eczema vaccinatum with methisazone (Bauer, 1965)

Assessment	Case No.	Age	Sex	Outcome	AVGG§ Dose	AVGG§ Effect	Methisazone dose (mg/kg) Initial	Methisazone dose (mg/kg) Total	Duration of treatment (days)	Side effects
Effective	1	7/12	M	Recovered	Given	None	125	876	7	Diarrhoea, Vomiting
	2	1	F	Recovered	—	—	200	600	2	Vomiting
	3	2	M	Recovered	—	?	200	600	2	Vomiting
0/12*	4	3	M	Recovered	1 g	—	35	700	5	Vomiting
	5	8/12	M	Recovered	—	—	200	600	2	None
	6	4	M	Recovered	"3 doses"	None	125	875	7	None
152/869†	7	52	F	Recovered	—	—	70	350	5	Nausea
	8	2	M	Recovered	—	—	150	450	3	Vomiting
	9	2½	M	Recovered	40 ml	None	200	520	2	None
3·75 days‡	10	2	M	Recovered	2 g	?	250	750	2	None
	11	1½	M	Recovered	—	—	200	560	3	None
	12	Adult	M	Recovered	4 g	None	80	400	5	None
Doubtful	13	27	F	Recovered	4 g	?	60	300	4	None
	14	1–1/12	M	Recovered	0·5 g	?	100	400	4	Diarrhoea, Vomiting
1/6	15	28	F	Recovered	6 g	?	85	340	4	None
63/324	16	1½	M	Died	1 g	None	22	66	1½	None
	17	1	M	Recovered	1·5 g	None	75	600	5	None
4·08 days	18	7	M	Recovered	—	—	40	240	6	None
Ineffective	19	10/12	F	Died	1 g	None	15	45	3	None
3/4	20	4	F	Recovered	—	—	60	420	7	Diarrhoea
48/231	21	11/12	F	Died	—	—	70	210	3	None
4·50 days	22	35	F	Died	2 g	None	50	250	5	None

* Deaths/total number of cases
† Mean initial dose/mean total dose (mg/kg)
‡ Mean duration of treatment with methisazone
§ Antivaccinial gamma globulin

Mainwaring (1962) observed rapid recovery in a severe case treated with
0·75 g daily to a total of 6 g, but antivaccinial gamma-globulin had also
been given, and it was uncertain which agent was responsible for recovery.
Success in treatment was also reported by Webb *et al.* (1965), Marsh and
Mitchell (1965), Adels and Oppé (1966) and Jaroszyńska-Weinberger
(1970). There was considerable variation in the doses used, and an analysis
of these and a number of unpublished cases is given in Table 7.2 (Bauer,
(1965). In this series mortality affords an objective criterion of clinical
efficacy, and it is evident that fairly high doses must be given. The recommended course of treatment is an initial dose of 200 mg/kg, followed by
50 mg/kg at 6-hourly intervals to a total dose of 600 mg/kg. If a dose is lost
by vomiting it should be repeated.

Vaccinia gangrenosa

In primary smallpox vaccination the course of the infection is brought to a
halt by humoral· and cell-mediated immunity. In certain cases these
mechanisms may be impaired or absent, as in congenital agammaglobulinaemia and hypogammaglobulinaemia, and in patients undergoing·
immunosuppressive treatment. In such cases the lesion fails to heal and
undergoes indefinite enlargement. Metastatic lesions arise as a result of
viraemia and can occur in almost any organ of the body. These secondary
lesions also enlarge, until the destruction of tissues which results becomes
incompatible with survival. The condition is known as vaccinia gangrenosa,
and is invariably fatal in the absence of treatment. Recovery is therefore
effective proof of the efficacy of treatment.

Kempe (1960) reported that cure could be obtained with antivaccinial
gamma globulin in some cases. Bauer (1965) analysed the results obtained
in ten cases of vaccinia gangrenosa treated with methisazone (Table 7.3).
Virological cure was obtained in five cases, two of which had failed to
respond to antivaccinial gamma globulin. The five cases in which treatment
was unsuccessful include those described by Connolly *et al.* (1962), Flewett
and Ker (1963) and White (1963). If treatment is to succeed it is evident
that high doses are required, and the dosage schedule recommended is the
same as that used in the treatment of eczema vaccinatum, namely, a
loading dose of 200 mg/kg, followed by 50 mg/kg every 6 hours to a total
dose of 600 mg/kg. The course may be repeated if necessary after an interval of 7 days. After virological cure the patient is still left with the underlying defects which gave rise to the condition in the first place.

Successful treatment of vaccinia gangrenosa with methisazone has also
been reported by Van Rooyen *et al.* (1967), Brainerd *et al.* (1967) and Hansson *et al.* (1966). The result obtained in the latter case is illustrated in
Figure 7.1. Kempe *et al.* (1967) obtained virological cure with methisazone
in five of nine patients with extensive vaccinia gangrenosa which had failed

Table 7.3　Treatment of vaccinia gangrenosa with methisazone (Bauer, 1965)

Outcome*	Case No.	Age	Sex	Antivaccinial gamma-globulin		Methisazone dose (mg/kg)		Duration of methisazone treatment (days)	side effects
				Dose	Effect	Initial	Total		
Recovered	1	21	M	2 g	None	83	830	10	None
116/638†	2	40	F	—	—	80	560	7	None
	3	7/12	F	—	—	200	600	3	None
6·2 days‡	4	30	M	2 g	?	18	500	7	None
	5	1–1/12	F	90 ml	None	200	700	4	None
Died	6	10/12	F	188 ml	None	25	1500	14	'Toxic'
	7	1–1/12	M	Given	None	30	730	7	None
84/791	8	4/12	M	2 g	None	166	1166	7	Diarrhoea
	9	6/12	M	14 g	None	60	180	2	None
6·5 days	10	7/12	M	2 g	None	125	875	7	Diarrhoea

* Of vaccinial infection
† Mean initial and total doses
‡ Mean duration of treatment with methisazone

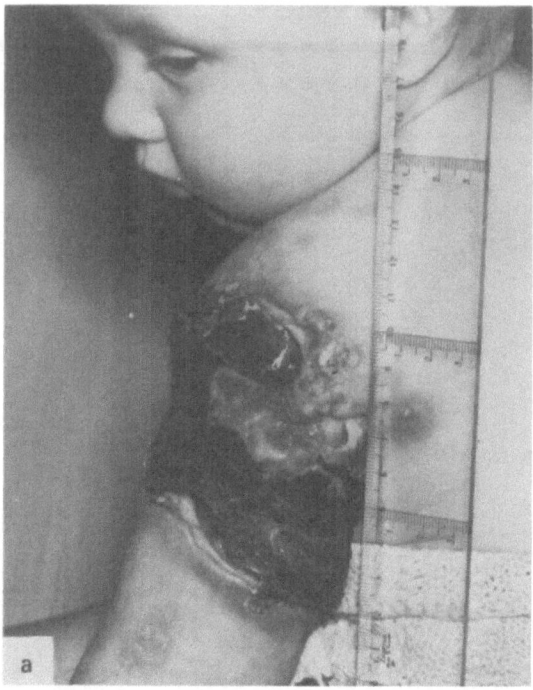

Figure 7.1 Treatment of vaccinia gangrenosa with methisazone. (a) Condition
before treatment. (b) Final result after two courses of methisazone
and skin grafting. [Hansson *et al.*, 1966.]

to respond to immunotherapy. He suggested that the patients who can be
cured are those which have a single immunological defect, such as Bruton-
type agammaglobulinaemia, or a defect resulting from immunosuppressive
treatment and underlying malignant conditions. In patients with multiple
defects (Swiss-type agammaglobulinaemia) treatment is ineffective.

Ocular vaccinia

Infection of the eye and orbit results from accidental transfer of the virus
from the site of vaccination. The various clinical forms of ocular vaccinia
have been described by Jones and Al-Hussaini (1963).

The commonest feature is oedema and cellulitis of the eyelids. The series
reported by Jaroszyńska-Weinberger included seven such cases, which
responded well to treatment with methisazone in a total dose of 3·2–11·0 g.
Conjunctivitis or blepharoconjunctivitis may also be present.

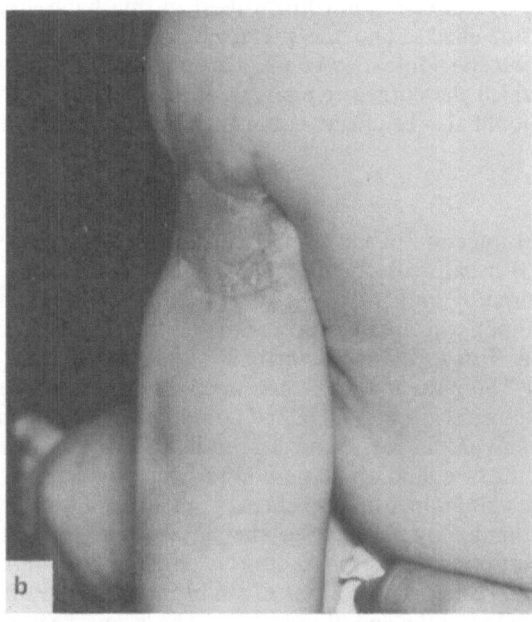

The most serious form of ocular vaccinia is keratitis. This may be a mild and self-limiting ulceration of the epithelium, or an infection of the stroma leading to disciform opacity or perforation of the cornea. Jones and Al-Hussaini (1963) found that experimental vaccinial keratitis in rabbits could be cured by the application of eye drops containing 0·1% idoxuridine, and suggested that this treatment should be used in man. A case treated with idoxuridine was described by Jack and Sorenson (1963). The patient was a boy aged 2 years who had vaccinial lesions on the right forefinger, ear, near the site of primary vaccination and the right eyelid. There was conjunctivitis in the left eye and a linear ulceration of the cornea. Eye drops of idoxuridine were given every hour during the day and every 2 hours during the night. Marked improvement occurred within 48 hours and treatment was discontinued after 5 days. The condition recurred 2 weeks later but responded to a further course of treatment with idoxuridine. Gordon and Advocate (1965) treated a case of vaccinial blepharokeratitis with cytarabine. The patient had rubbed the vaccination site and also his right eye. On examination 14 days after vaccination there was a vaccinial lesion on the right lower lid and an ulcer on the adjacent part of the cornea. There was a marked anterior chamber reaction and the cornea was oedematous and anaesthetic. The pupil was dilated with 1% cyclopentolate and

1% cytarabine ointment was instilled every 2 hours. On the 8th day of treatment the condition was largely resolved and the patient was then lost to treatment. Cytarabine is not much used in ophthalmology on account of its toxic side-effects, and the preferred method of treatment is with idoxuridine, which should also be used prophylactically in cases of ocular vaccinia in which the cornea remains unaffected. Trifluorothymidine and vidarabine should also be effective, but there are no reports of their use.

Cowpox

Cowpox is an infection of the skin of cattle with a poxvirus belonging to subgroup 1. It usually affects the teats and udders of cows. The infection may be acquired by milkers, and gives rise to lesions resembling those of primary smallpox vaccination occurring on the hands, arms and face. McMath and Wilson (1965) described a case which was treated with methisazone. The patient was a man aged 28 who had been vaccinated against smallpox in childhood. He had contact with a calf, and 6 days later developed nausea, dizziness and lassitude. Three days later swellings appeared on the face, and on admission to hospital 8 days later there were lesions on the chin, upper lip and cheek, consisting of crusts with purulent margins surrounded by a zone of erythema. Methisazone was given on the 10th day of eruption in an initial dose of 200 mg/kg (14 g), followed by 50 mg/kg (3·5 g) 6-hourly for 8 doses. Vomiting occurred on several occasions but ceased after the end of treatment. The lesions began to dry up, but it was uncertain whether this was due to treatment with methisazone, which was not given until very late in the course of the eruption.

Cowpox virus is much less sensitive to chemotherapy than vaccinia (Bauer, 1961), and it is doubtful whether any effect could be expected from treatment with methisazone.

There are no reports of the specific treatment of orf, paravaccinia or bovine papular stomatitis, and it is not known whether these viruses are inhibited by the chemotherapeutic agents available at present.

Smallpox

Smallpox occurs in two clinical variants, variola major and variola minor or alastrim. Variola major is typically a very severe illness with a mortality around 30%, although its course may be very much modified by immunity due to previous vaccination. Alastrim used to occur in South America, mainly in Brazil, and was a much milder illness with a mortality of 0·5%. The viruses of variola major and alastrim are closely similar but can be distinguished by their properties in laboratory tests.

The initial site of multiplication of the virus after entry into the body is not known, but is probably somewhere in the respiratory tract. After an incubation period of about 12 days the virus escapes into the blood stream, and further cycles of multiplication take place in the dermis and also in the internal organs. The viraemic phase is associated with a severe prodromal illness with severe backache, headache and prostration. The severity of the clinical course is directly related to the amount of virus liberated into the bloodstream during viraemia, and also to the degree of pre-existing immunity. During the incubation period the multiplication of the virus may be arrested by the administration of methisazone.

Smallpox affords a particularly favourable opportunity for carrying out chemoprophylaxis, since contacts are usually aware that they have been exposed to a patient suffering from smallpox. In a prophylactic trial of methisazone carried out by Bauer *et al.* (1964) in contacts of variola major in Madras, the drug was given according to four dosage schedules. The treated contacts were observed over a period of 14–16 days and cases of smallpox occurring among them were recorded and compared with those occurring among 2414 untreated contacts from the same environment. No cases of smallpox occurred during the observation period among 57 contacts treated with 3 g twice daily for 4 days. One case occurred among 384 given $1\frac{1}{2}$ g twice daily for 4 days, none among 584 given 2 doses of 3 g with an interval of 8–10 hours or so, and 5 among 1137 who took a single dose of 3 g. In the group of untreated contacts 102 cases occurred. The overall results of the trial are given in Table 7.4. The reduction in the incidence of contact cases due to treatment was around 6-fold.

Prophylactic trials against variola major were subsequently carried out by Rao *et al.* (1969 a) and Heiner *et al.* (1971), and against alastrim by Ribeiro do Valle *et al.* (1965), and an analysis of the results in comparison with the first trial is given in Table 7.5. In a trial confined to persons, mainly

Table 7.4 **Effect of methisazone treatment on the incidence of contact cases of smallpox**

Group	Treatment	Contacts	Cases	Deaths	Case incidence (%)
1	Treated, all dose levels	2287	6	2	0·26
2	Not completed	323	12	2	3·71
1 + 2	Total treated	2610	18	4	0·69
3	Not taken	150	11	3	7·33
4	Not offered	2665	105	18	3·94
3 + 4	Total untreated	2815	116	21	4·12

Table 7.5   Prophylaxis of variola with methisazone

| | | Cases/Contacts | | Significance of |
	Trial	Treated	Untreated	case reduction (p)
I	Bauer et al., 1967	18/2610	113/2710	< 0·001
II	Rao et al., 1969	2/17	8/19	0·047
III	R. do Valle et al., 1965	8/384	42/520	< 0·001
IVa	Heiner et al., 1971	1*/45	6/42	0·082
IVb	Heiner et al., 1971	2/111	3/109	0·491
IVc	Heiner et al., 1971	4/106	5/147	0·566

* Treated after onset of smallpox

infants, who had never been vaccinated. Rao et al. (1969 a) observed a 4-fold reduction in incidence. Methisazone was given in an adult dose of 5 g daily for 3 days, with a proportionate reduction in children and infants. They claimed that the result was not significant, but the reduction just attains significance, with $p = 0·047$. Ribeiro do Valle et al. (1965) obtained a 4-fold reduction in the incidence of contact cases of alastrim which was highly significant. Treatment consisted of one or two doses of 3 g.

The studies of Heiner et al. (1971) comprised three separate trials. The first was carried out in 1964 with two doses of 3 g given with an interval of 4–6 hours. There were six cases among 52 untreated contacts and one among 45 given methisazone. The reduction in case incidence verged on significance. In one case in the treated group methisazone was given after the onset of clinical smallpox, and this case should properly have been excluded from a prophylactic trial. If this is done the difference attains significance, with $p = 0·021$. The remaining two trials were carried out between 1965 and 1970, and no prophylactic effect was obtained. This inconsistent result may have been due to the fact that the treatment was changed to a single dose of 6 g. Such a large dose would be expected to cause vomiting, with loss of drug and chemoprophylactic effect. The incidence of vomiting recorded was extremely low, which makes it likely that contacts who vomited refused to admit it. The trials of Heiner et al. (1971) may be taken as indicating that two doses of 3 g are effective in the prophylaxis of smallpox whereas a single dose of 6 g is not.

Methisazone has been used on two occasions in epidemic conditions. Ferguson (1964) reported four cases of smallpox among 43 contacts given methisazone in the course of an epidemic of smallpox in South Africa in 1964. He described his work as a well-controlled investigation, but there was in fact no control group, and Heiner et al. (1971) interpreted these cases as individual instances of drug failure.

Methisazone was also used in Yugoslavia during an epidemic of smallpox in 1971 (Janković et al., 1972). It was given to 232 hospital staff members in direct contact with smallpox patients. Smallpox developed in five contacts; three of these had vomited the drug, and the remaining two were treated on the presumed 14th day of the incubation period. There was no control group, but observations were maintained on 515 indirect contacts, mostly staff and patients in other wards of the same hospital; 28 of these developed smallpox. No firm conclusion could be drawn from these observations, but the authors considered that methisazone had proved satisfactory when used in epidemic conditions, and stated that the treatment of choice for close contacts should be vaccination and treatment with methisazone.

The same recommendation was made in the Report of the Committee of Enquiry into the Smallpox Outbreak in London in March and April 1973 (1974), who advised in addition that contacts should be given antivaccinial gamma globulin.

Methisazone has not proved effective in the treatment of smallpox. Rao et al. (1969 b) carried out a double-blind trial of methisazone against a placebo. Methisazone was given to 208 patients in a dose of 3 g every 6 hours to a total of 12 doses, the dose being reduced proportionately in infants and children, and 215 patients were similarly treated with a placebo. Analysis of the results according to mortality and duration of illness showed that treatment with methisazone was ineffective. The lack of effect could have been due to the late stage at which most patients were admitted to hospital, and it is also possible that drug levels in the dermis were inadequate on account of the omission of a loading dose.

Hossain et al. (1972) treated nine smallpox patients with cytarabine during an outbreak in Bangladesh. The investigation was uncontrolled. Cytarabine was given subcutaneously or intravenously in an initial dose of 60 mg/m^2, followed by 60 mg/m^2 each day for 2 days, and then 30 mg/m^2 per day for 2 further days. Continuous medical supervision was not possible and the drug was therefore administered by continuous subcutaneous infusion. Eight patients had never been vaccinated, and the remaining patient was given primary vaccination on the day of onset of illness. Treatment was begun on the 5th–9th day of illness. There was a marked clinical improvement in all nine patients; eight recovered satisfactorily but one died of pneumonia. Normally three or four deaths would have been expected in a similar group of unvaccinated patients. No evidence of toxicity was noted with the dosage schedule employed, and the authors suggested that the apparently favourable results warranted the carrying out of a double-blind controlled trial. This was subsequently done by Dennis et al. (1974) in a trial carried out in Ethiopia. Cytarabine was given to nine patients in single intravenous doses of 100 mg/m^2 on each of 4 successive days, and nine patients were treated similarly with a placebo.

The treatment had no effect on the course of the disease. The observations of Hossain *et al.* (1972) could therefore not be confirmed, but the drug was administered by a different dosage schedule so that the two investigations are not strictly comparable.

Vidarabine has also been found ineffective in the treatment of smallpox (Koplan *et al.*, 1975).

GENERAL READING

Bauer, D. J. (1962). Thiosemicarbazones. In: D. J. Bauer (ed.) *Chemotherapy of Virus Diseases*, vol. 1, 35–113. (Oxford: Pergamon Press)

Dixon, C. W. (1962). *Smallpox*. (London: J. and A. Churchill Ltd.)

Chapter 8

Chemotherapy of myxovirus and papovavirus infections

MYXOVIRUSES

The classification of the myxoviruses is shown in Table 8.1. The only diseases in which specific chemotherapy has been used so far are influenza and subacute sclerosing panencephalitis. There are three types of influenza virus, A, B and C. The first two types are indistinguishable morphologically, whereas type C differs in some respects but is nevertheless classed as an influenza virus. Types A and B give rise to epidemics, with type A predominating, and type C causes sporadic respiratory infections. Type A is divided into subtypes of virus of human, porcine, equine and avian origin. Strains of mammalian and avian subtypes are of importance in human medicine, since they may undergo recombination with strains of human

Table 8.1   Classification of the myxoviruses

Genus	Example
Orthomyxovirus	Influenza A Influenza B Influenza C
Paramyxovirus	Mumps Parainfluenza types 1–4 Newcastle disease Measles Distemper Rinderpest Respiratory syncytial virus

subtype, giving rise to new strains to which the general population lacks immunity, as a result of which an epidemic may arise. This change of antigenic properties is known as antigenic shift, and occurs every 10 years or so, when a new subtype attains world-wide prevalence. A pandemic thus arises, and declines only when sufficient herd immunity has developed among the population. The new strain may subsequently undergo minor changes in antigenicity known as antigenic drift. Herd immunity is then less effective, and minor epidemics can arise.

An outline classification of the human subtypes of influenza A and their years of prevalence is given in Table 8.2.

Table 8.2   Outline classification of human subtypes of influenza A

Years of prevalence	Subtype	Antigenic composition
1933–1946	A0	H0N1
1946–1957	A1	H1N1·
1957–1967	A2 (Asian)	H2N2
1967 continuing	A3 (Hong Kong)	H3N2

Influenza

Influenza is transmitted by droplet infection. The virus multiplies in the ciliated cells of the respiratory mucosa, which is largely destroyed and desquamated. A process of repair then sets in, and the ciliated epithelium is largely restored after a week or so. The incubation period is about 2 days, by which time the virus has multiplied to a sufficient extent to cause clinical illness. The onset is acute, with fever, malaise, headache and muscle pain, with recovery setting in after 3 or 4 days. Bronchopneumonia may develop as a complication, and the disease may be fatal in elderly persons and in patients with chronic disease of the heart or respiratory system.

The multiplication of type A strains of influenza virus is inhibited by amantadine hydrochloride, and this compound has been shown to be effective in the prophylaxis and treatment of influenza in volunteer studies and also in epidemic conditions.

A prophylactic effect in volunteers was first reported by Jackson et al. (1964). College students who lacked antibody to influenza were divided into treatment and control groups of around 100 subjects each and infected experimentally with the Asian strain of influenza virus. Amantadine hydrochloride was given to the volunteers in the treatment group in a dose of 100 mg by mouth every 12 hours for 6 days. In one-half of the group treatment was begun 18 hours before infection in order to detect a prophylactic effect, and in the other half treatment was begun 4 hours after infection for observation of a therapeutic effect. The subjects in the control group were treated similarly with a placebo. The effect of treatment was assessed on the basis of occurrence of clinical illness and excretion of virus. Infection developed in 37% of those given pretreatment, and in 66% of the comparable members of the control group. The reduction in incidence was highly significant ($p < 0.01$), and indicated that amantadine hydrochloride had a prophylactic effect. No difference indicative of a therapeutic effect could be observed in the groups in which treatment was begun after infection. Tyrrell et al. (1965) were unable to observe any prophylactic effect in a small trial carried out on 17 volunteers, but the findings of Jackson et al. (1964) have been confirmed in a number of subsequent trials, three of which were carried out with Hong Kong strains. The results are summarized in Table 8.3.

Table 8.3   Studies in volunteers of the prophylactic effect of amantadine hydrochloride against influenza A infection

Reference	Type of virus	Treated*	Placebo*	Significant effect
Schiff et al. (1966)	Asian	0/9	4/9	+
Togo et al. (1968)	Asian	8/29	18/29	+
Smorodintsev et al. (1970 b)	Asian	56/122	106/166	+
Smorodintsev et al. (1970 b)	Hong Kong	6/17	13/16	+
Likar (1970)	Hong Kong	2/66	12/75	+
Smorodintsev et al. (1972)	Asian and Hong Kong	121/307	227/335	+

* No. of cases of infection/no. in group

Amantadine hydrochloride also has a prophylactic effect in natural infections with influenza virus. Wendel et al. (1966) carried out a double-blind trial among volunteers during an epidemic of Asian influenza in a prison. Amantadine hydrochloride was given by mouth in a daily dose of 200 mg for 10–13 days. During the period of treatment five of 439 (1.1%) developed influenza, compared with 15 of 355 (4.2%) who had been given a placebo. The reduction in case incidence was 3.7-fold and was significant ($p < 0.01$).

A much more extended trial was carried out by Finklea et al. (1967). Amantadine hydrochloride and placebo were administered under double-blind conditions to mentally retarded children living in a residential school. The dose ranged from 60 mg to 100 mg according to age, equivalent to 1·0–2·5 mg/kg. Treatment was continued for a period of 4 months in late winter and spring. A rise in titre of antibody to Asian influenza of 4-fold or more, indicative of infection, occurred in one of 104 children in the treated group and in 11 of 133 given placebo. The reduction in case incidence was 8·8-fold.

Galbraith *et al*. (1969 a) carried out a prophylactic trial of amantadine hydrochloride among the household contacts of index cases of Asian influenza. The trial was carried out on a co-operative basis in 22 general practices distributed over Great Britain. Amantadine hydrochloride was given in a dose of 100 mg every 12 hours to all the members of a particular family, and other families were treated similarly with a placebo. Over an observation period of 10 days two cases of clinical influenza occurred among 55 persons given amantadine hydrochloride, compared with 12 among 85 given placebo. The reduction in case incidence was 4·6-fold. When the results were confined to cases in which serological confirmation was obtained there were no cases among 48 given amantadine hydrochloride and 10 among 69 given placebo. The reduction in case incidence was significant (p 0·01–0·05).

The same authors (Galbraith *et al*., 1969 b) carried out a similar study during a subsequent epidemic of Hong Kong influenza. No prophylactic effect could be observed, since the case incidence was 5/44 in the treated group and 6/42 in the placebo group. Since the Hong Kong strain is sensitive to amantadine hydrochloride the authors attributed the negative result to the general absence of antibody to the new strain among the study population. However, other investigators have demonstrated a prophylactic effect against Hong Kong influenza in trials involving much larger numbers of subjects.

A general summary of prophylactic trials of amantadine hydrochloride in natural influenza is given in Table 8.4.

Amantadine hydrochloride has also been tried in the treatment of influenza. Wendel *et al*. (1966) found no evidence of therapeutic effect in a small-scale trial. Wingfield *et al*. (1969) carried out a double-blind trial in prisoners during an outbreak of Asian influenza in 1968. Amantadine hydrochloride was given in a dose of 100 mg twice daily for 10 days to 23 subjects and 47 were treated similarly with a placebo. There was a significant reduction in the duration of illness in the treated group. A similar effect was observed by Knight *et al*. (1970) in influenza caused by virus of Hong Kong type.

A summary of reports on the therapeutic use of amantadine hydrochloride is given in Table 8.5.

Amantadine hydrochloride has not been licensed for use against Hong Kong influenza in the USA. The indications for which it is recommended

Table 8.4 Prophylactic trials of amantadine hydrochloride in natural outbreaks of influenza

Reference	Type of virus	Treated	Placebo	Reduction†	Significance
Wendel et al., 1966	Asian	5*/439*	15*/355*	3·7	+
Quilligan et al., 1966	Asian	12/126	13/43	3·3	+
Finklea et al., 1967	Asian	1/104	11/133	8·8	+
Callmander and Hellgren, 1968	Asian	33/48	31/48	—	−
Galbraith et al., 1969 a	Asian	2/55	12/85	4·6	+
Farkas, 1969	Hong Kong	59/2530	220/2210	4·5	+
Galbraith, 1969	Hong Kong	5/44	6/42	—	−
Smorodintsev et al., 1970 a	Hong Kong	156/3885	192/2498	2·0	+
Maté et al., 1970	Hong Kong	173/2435	129/2127	—	−
Nafta et al., 1970	Hong Kong	2/112	20/103	10·9	+
Oker-Blom et al., 1970	Hong Kong	27/192	57/199	2·0	+
Bloomfield et al., 1970	Hong Kong	8/157	15/140	2·0	±
O'Donoghue et al., 1972	Hong Kong	0/50	7/61	>7·0	+

* No. of contact cases/no. in group
† Case incidence in control group/case incidence in treated group

Table 8.5 Trials of the therapeutic effect of amantadine hydrochloride in natural influenza infections

Reference	Type	Amantadine HCl	Placebo	Significant effect
Wendel et al., 1966	Asian	30	25	−
Hornick et al., 1969	Asian	94	103	+
Wingfield et al., 1969	Asian	23	47	+
Baker et al., 1969	Asian	14	13	+
Togo et al., 1970	Asian	54	48	±
Kitamoto et al., 1970	Asian	182	173	+
Knight et al., 1970	Hong Kong	13	16	+
Kitamoto et al., 1970	Hong Kong	114	112	+
Galbraith et al., 1971	Hong Kong	72	81	+
Maté et al., 1971	Hong Kong	79	97	+
Kitamoto, 1971	Hong Kong	79	76	+

in Great Britain include the treatment of influenza in patients in whom complications may be expected, and prophylactic use in persons with chronic respiratory diseases or debilitating conditions.

Subacute sclerosing panencephalitis (SSPE)

Subacute sclerosing panencephalitis is a rare disease, usually of children and adolescents, which develops some years after an attack of measles with an intervening period of normal health. The first description of the condition was given by Dawson (1933), who observed eosinophil inclusions in the nuclei and cytoplasm of neurons and glia cells in autopsy material from a patient who had died from a disease diagnosed as encephalitis lethargica. Inclusions could not be found in material from three other cases, and Dawson therefore concluded that encephalitis lethargica was not a single clinical entity, and that the case which he observed was due to a virus infection.

The clinical course has been described by Foley and Williams (1953) and Metz et al. (1964). The illness begins with disturbances in behaviour, intellectual deterioration, petit mal attacks, episodes of akinesia and myoclonus leading to profound dementia and death after 6–12 months.

Horta-Barbosa et al. (1969) reported the isolation of a virus from brain biopsy material. This was cultured in the presence of HeLa or HEp-2 cells, which developed a cytopathic effect consisting of syncytia and giant cells. The agent responsible for the effect could be transferred to HeLa cell cultures and reacted positively with measles antisera in immunofluorescence tests. The authors considered that the agent was measles or a virus very similar to it.

Payne et al. (1969) co-cultivated brain biopsy material with a line of African green monkey kidney cells and also obtained an agent which they identified as measles virus. In similar work Barbanti-Brodano et al. (1970) isolated an agent with the general properties of measles virus from two of three brain biopsy specimens, but found that it differed from measles in the range of cell types which it was able to infect. Differences were also apparent in electron microscope studies carried out by Oyanagi et al. (1971). Both measles and the SSPE agent produced nucleocapsids in the form of smooth and rough filaments, but there were marked differences in the amounts found in the nuclei and cytoplasm. In measles the granular filaments lay beneath the plasma membrane and became incorporated into particles which budded from the surface, but no such budding was observed with the SSPE agent. The authors pointed out that these differences were only quantitative, but were nevertheless sufficient to distinguish SSPE from measles.

Ter Meulen et al. (1972 b) observed differences in sentivity to 6-azauridine, a synthetic nucleoside which was found to inhibit the multiplication of measles virus by Leonard et al. (1971). In plaque inhibition tests carried out in a continuous line of African green monkey kidney cells two strains of SSPE virus were completely inhibited by a concentration of 10 µg/ml in the overlay, whereas two strains of measles virus required 100 µg/ml for complete inhibition.

It therefore appears that subacute sclerosing panencephalitis is due to infection with a virus closely resembling but distinct from measles. The prolonged nature of the clinical course may result from the fact that the infected cells do not release virions, so that the virus can spread only to cells in immediate contact. An immune mechanism also plays a part in the pathogenesis of the disease.

Subacute sclerosing panencephalitis may also occur as an apparent complication of infectious mononucleosis. Feorino *et al.* (1975) observed a case in which the two conditions occurred simultaneously, and two others in which subacute sclerosing panencephalitis supervened upon pre-existing infectious mononucleosis. It is conceivable that a virus of the herpes group may be the causative agent in some cases.

Amantadine hydrochloride has been used in the treatment of subacute sclerosing panencephalitis, although there is no evidence that it can inhibit the multiplication of measles virus. Haslam *et al.* (1969) considered that they had obtained some favourable results in the treatment of six cases. Amantadine hydrochloride was given three times daily in a dose of 50 mg (6–10 mg/kg) to three patients. The daily dosage schedules in the remaining three patients were three doses of 50 mg (5 mg/kg), three doses of 200 mg (10 mg/kg) and two doses of 100 mg (3·5 mg/kg). The duration of treatment ranged from 40 days to more than 23 months. In five patients a tendency for the disease to become stabilized was observed, although there was no reversal of established symptoms. These patients were all alive at the time of publication, with survival times ranging up to 4½ years after the onset of illness, and the authors considered that this was additional evidence that the treatment had had some effect. They nevertheless stressed that much additional evidence will be required before the effect of amantadine hydrochloride in the treatment of subacute sclerosing panencephalitis can be satisfactorily evaluated.

There are no reports on the use of 6-azauridine, a specific inhibitor of the multiplication of measles virus. It has been used in the treatment of leukaemia in man, and its general properties and clinical pharmacology have been described by Handschuhmacher *et al.* (1962). The dosage used in the treatment of leukaemia has ranged from 10 to 20 g daily given intravenously over a period of 14–21 days. It seems worthy of trial in subacute sclerosing panencephalitis, since the treatment would at least have a rational basis. There are no preparations available for clinical use, but the compound is available commercially.

PAPOVAVIRUSES

The papovavirus family contains two genera, papillomavirus and polyomavirus. An abbreviated classification is given in Table 8.6. The members of the family of importance in clinical medicine are warts virus and papovaviruses, including SV40, which may be the aetiological agents of progressive multifocal leukoencephalopathy. SV40, a polyomavirus of rhesus monkeys, has also aroused clinical attention as a contaminant in early batches of poliomyelitis vaccine, as it has been shown to cause tumours in hamsters.

Table 8.6   Classification of the papovaviruses

Genus	Example
Papillomavirus	Warts
	Rabbit papilloma
	Papillomatosis of domestic and other animals
Polyomavirus	Progressive multifocal leukoencephalopathy
	SV40 virus
	Mouse polyoma

Warts

There is a single report on the use of idoxuridine in the treatment of warts, which is included here for the sake of completeness. Manilla *et al.* (1965) found that warts could be cleared by repeated application of 0·5% idoxuridine solution mixed with a hydrophilic cream base. There is some theoretical likelihood that the treatment could have been effective, since it is known that idoxuridine will inhibit the multiplication of SV40 and mouse polyomavirus. It seems that the use of idoxuridine and other antiviral nucleosides in the treatment of warts has not yet been properly evaluated. In the light of subsequent work on the treatment of cutaneous herpes and zoster it would seem preferable to try a 40% solution in dimethylsulphoxide.

Progressive multifocal leukoencephalopathy

This condition was first recognized by Åström *et al.* (1958), who described an unusual neurological disorder occurring in two patients with chronic lymphatic leukaemia and one with Hodgkin's disease. It bore no relation to any recognized complication of these conditions and was considered to

be a new disease entity. The main features of the condition were the progressive development of weakness, ataxia, hemiparesis, confusion, dis-
orientation, incontinence, diplopia and aphasia, leading to blindness and
death after 10 weeks to 4 months from the onset. In sections of the brain
there were many foci of perivascular destruction of myelin sheaths, with
pleomorphic microglia and hypertrophy and proliferation of astrocytes.
The lesions enlarged and became confluent. A striking feature was the
presence of cells with deeply staining nuclei of unknown origin. The
disease is a chronic progressive condition which develops by the formation
of new lesions and the enlargement of existing ones. It does not always
progress so rapidly, since Hedley-White et al. (1966) observed a patient
who went into remission and died after 5 years.

ZuRhein and Chou (1965) carried out an electron microscope examination of thin sections of brain material taken at autopsy, and observed
virus-like particles in the nuclei of glial cells. They were usually abundant,
spherical, 33–36 nm in diameter and randomly arranged. Crystalloid
arrays were sometimes seen, associated with filamentous forms of $\frac{1}{2}$–$\frac{2}{3}$ the
diameter of the spherical particles, arranged in strands or whorls, or
occurring singly. The appearances suggested a virus of the papovavirus
group. Similar observations were made by Silverman and Rubinstein
(1965). They found large numbers of virus-like particles in the nuclei of the
oligodendroglial cells. They were round, 35–40 nm in diameter, but there
were also filaments 35–40 nm in width and 125–625 nm in length. Rarely
there were bundles of filamentous forms up to 1600 nm in length. Particles
were sometimes found in the cytoplasm. The appearances were consistent
with infection with a papovavirus, and confirmed the observations of
ZuRhein and Chou (1965). Similar findings were also reported by Conomy
et al. (1974), who considered that they were due to infection with SV40 or a
related polyomavirus.

These electron microscope observations have been supported by the
isolation of viruses from brain biopsy material from patients suffering
from progressive multifocal leukoencephalopathy. Padgett et al. (1971)
inoculated extracts of biopsy material into primary cultures of human glial
cells. After incubation for 10–12 days a cytopathic effect developed, consisting of mounds containing normal cells, cells in mitosis, necrotic cells
and cells with intranuclear inclusion bodies. In negatively stained extracts
of affected cells many virions with a mean diameter of 42·3 nm were seen.
In thin sections virions were seen in the nucleus and cytoplasm, often
arranged in paracrystalline arrays. The virus (SC virus) was similar in size
to SV40 and polyomavirus but smaller than human papilloma (wart) virus.
It did not cross-react with SV40 or polyoma antisera in immunofluorescence
studies. Weiner et al. (1972) cultured brain biopsy material through several
passages and then fused the cells with primary African green monkey kidney
cells with inactivated Sendai virus. Homogenates of the cells were then passed

into cultures of green monkey kidney cells and a vacuolating virus appeared after two passages. Electron microscopy of the cells showed the presence of particles 33–35 nm in diameter situated in the nuclei and along endoplasmic membranes. In neutralization and immunofluorescence tests the virus was indistinguishable from SV40.

The results of electron microscopy and virus isolation studies thus indicate that progressive multifocal leukoencephalopathy is a subacute infection with a polyomavirus, some strains of which are indistinguishable from SV40.

The multiplication of papovaviruses is inhibited by the antiviral nucleosides, and some progress has been made in the treatment of progressive multifocal leukoencephalopathy with these agents. Bauer *et al.* (1973) described a case in which this condition supervened in a 36-year-old man suffering from chronic lymphatic leukaemia of 4 years duration, who developed dysarthria, left spastic hemiparesis and cerebellar tremor of the left arm. The diagnosis was confirmed by biopsy of the right frontal lobe. The condition of the patient continued to deteriorate, and it was decided to attempt treatment with cytarabine. The drug was given intravenously in doses of 60 mg/m^2 daily for 6 days and 10 mg/m^2 daily by the intrathecal route for 2 days. Within 48 hours of beginnning treatment the mental state and hemiplegia improved and the patient became alert, orientated and continent. He was discharged with dysarthria and walking with the aid of a stick. The condition was unchanged at examination 6 months later.

Conomy *et al.* (1974) observed clinical improvement in a 54-year-old man who developed the disease after treatment for chronic lymphocytic leukaemia for 12 years. At the time of treatment the patient had paresis of the right arm and was comatose. Cytarabine was given intravenously in a dose of 100 mg. There was rapid improvement within the first 24 hours, and the patient became alert and was able to sit up in bed and feed himself. Between the 3rd and 14th days after beginning treatment cytarabine was given intravenously in a daily dose of 30 mg, and 100 mg was given intrathecally on the 5th, 8th and 11th days. The improvement was not sustained, however, and continuing deterioration of the neurological condition led to death 7 days after the end of cytarabine treatment. The authors concluded that the use of cytarabine in progressive multifocal leukoencephalopathy warranted further investigation.

There is one report of the unsuccessful use of idoxuridine (Tarsy *et al.*, 1973). A patient with sarcoidosis developed progressive multifocal leukoencephalopathy which was confirmed by light and electron microscopy of a brain biopsy specimen. Idoxuridine was injected into the right lateral ventricle in a dose of 2 mg/kg (123 mg) every 12 hours for 21 days. It was administered in a volume of 10 ml after removing a corresponding amount of cerebrospinal fluid. There was no evidence of toxicity, and no evidence of change in the neurological condition. Treatment was discontinued for a week and then resumed for a further 28 days. There was no improvement and the patient died from an intercurrent infection 2 months after beginning treatment. At autopsy there were the typical signs of progressive multifocal leukoencephalopathy, and also periventricular haemorrhagic necrosis and gliosis mainly localized around the occipital horns of both ventricles, which could have been due to local toxicity of idoxuridine resulting from inadequate mixing in the most dependent parts of the ventricular system. It was concluded that treatment with idoxuridine had been ineffective.

GENERAL READING

Francis, T. and Maassab, H. F. (1965). Influenza viruses. In: F. L. Horsfall and I. Tamm (eds.) *Viral and Rickettsial Infections of man.* 4th edn. 689–740. (London: Pitman Medical Publishing Co. and Philadelphia: J. B. Lippincott Co.)

Bibliography

Adamson, R. H., Ablashi, D. V., Armstrong, G. R. and Ellmore, N. W. (1972). Effect of cytosine arabinoside, adenine arabinoside, tilorone, and rifamycin SV on multiplication of *Herpesvirus saimiri* in vitro. *Antimicrobial Ag. Chemother.*, **1**, 82–83

Adels, B. R. and Oppé, T. E. (1966). Treatment of eczema vaccinatum with *N*-methylisatin beta-thiosemicarbazone. *Lancet*, **ii**, 18–20

Adlard, B. P. F., Dobbing, J. and Sands, J. (1975). A comparison of the effects of cytosine arabinoside and adenine arabinoside on some aspects of brain growth and development in the rat. *Br. J. Pharmacol.*, **54**, 33–39

Andrewes, C. H. and Pereira, H. G. (1972). *Viruses of Vertebrates*. 3rd ed. (London: Baillière Tindall)

Ansfield, F. J. and Ramirez, G. (1971). Phase I and II studies of 2′-deoxy-5-(trifluoromethyl)uridine (NSC-75520). *Cancer Chemother. Rep.*, **55**, 205–208

Appleyard, G., Hume, V. B. M. and Westwood, J. C. N. (1965). The effect of thiosemicarbazones on the growth of rabbit pox virus in tissue culture. *Ann. N.Y. Acad. Sci.*, **130**, Art. 1, 92–104

Appleyard, G. and Way, H. J. (1966). Thiosemicarbazone-resistant rabbitpox virus. *Br. J. Exp. Pathol.*, **47**, 144–151

Åström, K. E., Mancall, E. L. and Richardson, E. P. (1958). Progressive multifocal leuko-encephalopathy. A hitherto unrecognized complication of chronic lymphatic leukaemia and Hodgkin's disease. *Brain*, **81**, 93–111

Aurelian, L. (1973). Virions and antigens of herpes virus type 2 in cervical carcinoma. *Cancer Res.*, **33**, 1539–1547

Axon, A. (1972). The presentation of methisazone (an antiviral chemical) as a medicine for oral administration. *J. Mond. Pharm.*, **4**, 221–232

Bach, M. K. and Magee, W. E. (1962). Biochemical effects of isatin β-thiosemicarbazone on development of vaccinia virus. *Proc. Soc. Exp. Biol. Med.*, **110**, 565–567

Bader, J. P. (1965). The requirement for DNA synthesis in the growth of Rous sarcoma and Rous-associated viruses. *Virology*, **26**, 253–261.

Baguley, B. C. and Falkenhaug, E.-M. (1971). Plasma half-life of cytosine arabinoside (NSC-63878) in patients treated for acute myeloblastic leukaemia. *Cancer Chemother. Rep.*, **55**, 291–298

Baker, L. M., Paulshock, M. and Iezzoni, D. G. (1969). The therapeutic efficacy of Symmetrel (amantadine hydrochloride) in naturally occurring influenza A2 respiratory illness. *J. Am. Osteopath. Ass.*, **68**, 1244–1250

Bakhle, Y. S., Sears, M. L. and Prusoff, W. H. (1965). Radioactive idoxuridine administered into the conjunctival sac of the rabbit. A study of the metabolism and tissue distribution. *Arch. Ophthal.*, **73**, 248–252

Balode, V. and Gibadûlin, R. A. (1970). Effect of aminoadamantane on DNA synthesis and the mitotic cycle of chick embryo fibroblast primary culture cells. *Latv. P.S.R. Zinat. Akad. Vestis*,74– 82 [from *Chem. Abstr.*, **73**, 43547h, 1970]

Barbanti-Brodano, G., Oyanagi, S., Katz, M. and Koprowski, H. (1970). Presence of two different viral agents in brain cells of patients with subacute sclerosing panencephalitis. *Proc. Soc. Exp. Biol. Med.*, **134**, 230–236

Baringer, J. R. (1974). Recovery of herpes simplex virus from human sacral ganglions. *N. Engl. J. Med.*, **291**, 828–830

Baringer, J. R. and Swoveland, M. A. (1973). Recovery of herpes-simplex virus from human trigeminal ganglions. *N. Engl. J. Med.*, **288**, 648–650

Baron, M. and Wechsler, H. L. (1975). Low-dosage cytarabine therapy for herpes zoster with pneumonia. *Arch. Dermatol.*, **111**, 910–912

Barton, B. W. and Tobin, J. O'H. (1970). The effect of idoxuridine on the excretion of cytomegalovirus. *Ann. N. Y. Acad. Sci.*, **173**, Art. 1, 90–95

Barz, P. and Fritz, H. P. (1970). Spektroskopische Untersuchungen an Metall-komplexen des Antivirusmittels Isatin-3-thiosemicarbazon. *Z. Naturforsch.*, **25B**, 199–204

Bastian, F. O., Rabson, A. S., Yee, C. L. and Tralke, T. S. (1972). Herpesvirus hominis: isolation from human trigeminal ganglion. *Science*, **178**, 306–307

Baublis, J. V., Whitley, R. J., Ch'ien, L. T. and Alford, C. A. (1975). Treatment of cytomegalovirus infection in infants and adults. In: D. Pavan-Langston, R. A. Buchanan and C. A. Alford (eds.) *Adenine arabinoside: an antiviral agent*, pp. 247–260 (New York: Raven Press)

Bauer, D. J. (1955). The antiviral and synergic actions of isatin β-thiosemi-carbazone and certain phenoxypyrimidines in vaccinia infection in mice. *Br. J. Exp. Pathol.*, **36**, 105–114

Bauer, D. J. (1961). The zero effect dose (E_0) as an absolute numerical index of antiviral chemotherapeutic activity in the pox virus group. *Br. J. Exp. Pathol.*, **42**, 201–206

Bauer, D. J. (1965). Clinical experience with the antiviral drug Marboran (1-methylisatin 3-thiosemicarbazone). *Ann. N. Y. Acad. Sci.*, **130**, Art. 1, 110–117

Bauer, D. J. and Apostolov, K. (1966). Adenovirus multiplication: inhibition by methisazone. *Science*, **154**, 796–797

Bauer, D. J., Apostolov, K. and Selway, J. W. T. (1970). Activity of methisazone against RNA viruses. *Ann. N. Y. Acad. Sci.*, **173**, Art. 1, 314–319

Bauer, D. J. and Collins, P. (1974). [Unpublished work]

Bauer, D. J. and Sadler, P. W. (1960 a). The structure-activity relationships of the antiviral chemotherapeutic activity of isatin β-thiosemicarbazone. *Br. J. Pharmacol. Chemother.*, **15**, 101–110

Bauer, D. J. and Sadler, P. W. (1960 b). New antiviral agent active against smallpox infection. *Lancet*, **i**, 1110–1111

Bauer, D. J. and Sheffield, F. W. (1959). Antiviral chemotherapeutic activity of isatin β-thiosemicarbazone in mice infected with rabbitpox virus. *Nature*, **184,** 1496–1497

Bauer, D. J., St. Vincent, L., Kempe, C. H., Young, P. A. and Downie, A. W. (1969). Prophylaxis of smallpox with methisazone. *Am. J. Epidemiol.*, **90,** 130–145

Bauer, W. R., Turel, A. P. and Johnson, K. P. (1973). Progressive multifocal leukoencephalopathy and cytarabine. Remission with treatment. *J. Am. Med. Ass.*, **226,** 174–176

Benda, R. (1965). Attempt at inhibiting B virus (herpesvirus simiae) growth in rabbits by 5-iodo-2'-deoxyuridine. *Acta Virol.*, **9,** 556

Betts, R. F., Zaky, D. A., Douglas, R. G. and Royer, G. (1975). Ineffectiveness of subcutaneous cytosine arabinoside in localized herpes zoster. *Ann. Intern. Med.*, **82,** 778–783

Biantrate, P., Tognoni, G., Belvedere, G., Frigero, A., Rizzo, M. and Morselli, P. L. (1972). A gas chromatographic method for the determination of amantadine in human plasma. *J. Chromatogr.*, **74,** 31–34

Bleidner, W. E., Harmon, J. B., Hewes, W. E., Lynes, T. E. and Herrmann, E. C. (1965). Absorption, distribution and excretion of amantadine hydrochloride. *J. Pharmac. Exp. Ther.*, **150,** 484–490

Bloomfield, S. S., Gaffney, T. E. and Schiff, G. M. (1970). A design for the evaluation of antiviral drugs in human influenza. *Am. J. Epidemiol.*, **91,** 568–574

Bodey, G. P., Gottlieb, J., McCredie, K. B. and Freireich, E. J. (1974). Arabinosyl adenine (ara-A) as an antitumor agent. *Proc. Am. Ass. Cancer Res.*, **15,** Abstr. 514, 129

Bondarev, V. N., Torosov, T. M., Kessova, E. T., Piskareva, N. A. and Denisov, G. M. (1970). [Investigation of the prophylactic effect of methisazone in smallpox vaccination of children with relative contraindications. In Russian.] *Zh. Mikrobiol. Epidem. Immunobiol.*, No. 9, 12–26

Borondy, P. E., Mourer, D. R., Drach, J. C., Chang, T. and Glazko, A. J. (1973). Species differences in the metabolic disposition of 9-β-D-arabino-furanosyladenine (Vidarabine; ara-A). *Fed. Proc.*, Abstr. 3172, 776

Borsa, J., Whitmore, G. F., Valeriote, F. A., Collins, D. and Bruce, W. R. (1969). Studies on the persistence of methotrexate, cytosine arabinoside, and leucovorin in serum of mice. *J. Nat. Cancer Inst.*, **42,** 235–242

Boston Interhospital Virus Study Group (1975). Failure of high dose 5-iodo-2'-deoxyuridine in the therapy of herpes simplex virus encephalitis. Evidence of unacceptable toxicity. *N. Engl. J. Med.*, **292,** 599–605

Boyce, M. J. (1972). E.B. virus and the Guillain-Barré syndrome. *Lancet*, ii, 1028–1029

Brainerd, H. D., Hanna, L. and Jawetz, E. (1967). Methisazone in progressive vaccinia. *N. Engl. J. Med.*, **276,** 620–622

Bredt, A. B. and Mardiney, M. R. (1969). Effects of amantadine on the reactivity of human lymphocytes stimulated by allogeneic lymphocytes and phytohemagglutinin. *Transplantation*, **8,** 763–773

Breeden, C. J., Hall, T. C. and Tyler, H. R. (1966). Herpes simplex encephalitis treated with systemic 5-iodo-2'-deoxyuridine. *Ann. Intern. Med.*, **65,** 1050–1056

Bresnick, E. and Williams, S. S. (1967). Effects of 5-trifluoromethyldeoxyuridine upon deoxythymidine kinase. *Biochem. Pharmacol.*, **16**, 503–507

Brink, J. J. and LePage, G. A. (1964 a). Metabolic effects of 9-D- arabinosylpurines in ascites tumor cells. *Cancer Res.*, **24**, 312–318

Brink, J. J. and LePage, G. A. (1964 b). Metabolism and distribution of 9-β-D-arabinofuranosyladenine in mouse tissues. *Cancer Res.*, **24**, 1042–1049

Brown, R. S. and Bower, B. D. (1975). Neonatal disseminated herpes simplex virus infection with encephalitis treated with cytosine arabinoside. *Devel. Med. Child Neurol.*, **17**, 493–498

Brownlee, K. A. and Hamre, D. (1951). Studies on chemotherapy of vaccinia virus. I. An experimental design for testing antiviral agents. *J. Bacteriol.*, **61**, 127–134

Buthala, D. A. (1964). Cell culture studies on antiviral agents: I. Action of cytosine arabinoside and some comparisons with 5-iodo-2-deoxyuridine. *Proc. Soc. Exp. Biol. Med.*, **115**, 69–77

Cairns, J. E. (1964). Varicella of the cornea treated with 5-iodo-2'-deoxyuridine. *Br. J. Ophthal.*, **48**, 288–289

Calabresi, P. (1963). Current status of clinical investigations with 6-azauridine, 5-iodo-2'-deoxyuridine and related derivatives. *Cancer Res.*, **23**, 1260–1267

Calabresi, P., Cardoso, S. S., Finch, S. C., Kligerman, M. M., von Essen, C. F., Chu, M. Y. and Welch, A. D. (1961). Initial clinical studies with 5-iodo-2'-deoxyuridine. *Cancer Res.*, **21**, 550–559

Calabresi, P., Finch, S. C., Cardoso, S. S. and Welch, A. D. (1960). Preliminary clinical experience with 5-iododeoxyuridine. *Proc. Am. Ass. Cancer Res.*, 99

Callmander, E. and Hellgren, L. (1968). Amantadine hydrochloride as a prophylactic in respiratory infections. A double-blind investigation of its clinical use and serology. *J. Clin. Pharmacol.*, **8**, 186–189

Camiener, G. W. and Smith, C. G. (1965). Studies on the enzymatic deamination of cytosine arabinoside. I. Enzyme distribution and species specificity. *Biochem. Pharmacol.*, **14**, 1405–1416

Campbell, J. B., Maes, R. F., Wiktor, T. J. and Koprowski, H. (1968). The inhibition of rabies virus by arabinosyl cytosine. Studies on the mechanism and specificity of action. *Virology*, **34**, 701–708

Cangir, A., Sullivan, M. P., Sutow, W. W. and Taylor, G. (1967). Cytomegalovirus syndrome in children with acute leukaemia. Treatment with floxuridine. *J. Am. Med. Ass.*, **201**, 612–615

Carlton, C. A. and Kilbourne, E. (1952). Activation of latent herpes simplex by trigeminal sensory-root section. *N. Engl. J. Med.*, **246**, 172–176

Carp, R. I., Licursi, P. C., Merz, P. A. and Merz, G. S. (1972). Decreased percentage of polymorphonuclear neutrophils in mouse peripheral blood after inoculation with material from multiple sclerosis patients. *J. Exp. Med.*, **136**, 618–629

Carter, A. R. (1968). Cytomegalovirus disease presenting as hepatitis. *Br. Med. J.*, **3**, 786

Caunt, A. (1967). The effect of Marboran on the growth of varicella virus in tissue culture. *Vth International Congress of Chemotherapy*, Vienna, **4**, 313–317

Centifanto, Y. and Kaufman, H. E. (1965). Thymidine kinase in IDU resistance. *Proc. Soc. Exp. Biol. Med.*, **120**, 23–26

Cheatham, W. J., Weller, T. H., Dolan, T. F. and Dower, J. L. (1956). Varicella: report of two fatal cases with necropsy, virus isolation and serologic studies. *Am. J. Pathol.*, **32**, 1015–1035

Ch'ien, L. T., Cannon, N. J., Charamella, L. J., Dismukes, W. E., Whitley, R. J., Buchanan, R. A. and Alford, C. A. (1973 a). Effect of adenine arabinoside on severe *Herpesvirus hominis* infections in man. *J. Infect. Dis.*, **128**, 658–683

Ch'ien, L. T., Schabel, F. M. and Alford, C. A. (1973 b). Arabinosyl nucleosides and nucleotides. *Selective Inhibitors of Viral Functions*, W. A. Carter (ed.), 227–258 (Cleveland, Ohio: Cleveland Rubber Co. Press)

Ch'ien, L. T., Whitley, R. J., Nahmias, A. J., Lewin, E. B., Linnemann, C. C., Frenkel, L. D., Bellanti, J. A., Buchanan, R. A. and Alford, C. A. (1975). Antiviral chemotherapy and neonatal herpes simplex virus infection: a pilot study—experience with adenine arabinoside (ara-A). *Pediatrics*, **55**, 678–685

Chow, A. W., Foerster, J. and Hryniuk, W. (1970). Cytosine arabinoside therapy for herpesvirus infections. *Antimicrob. Ag. Chemother.*, 214–217

Chow, A. W., Ronald, A., Fiala, M., Hryniuk, W., Weil, M. L., St. Geme, J. and Guze, L. B. (1973). Cytosine arabinoside therapy for herpes simplex encephalitis—clinical experience with six patients. *Antimicrob. Ag. Chemother.*, **8**, 412–417

Chu, M.-Y. (1971). Incorporation of arabinosyl cytosine into 2-7S ribonucleic acid and cell death. *Biochem. Pharmacol.*, **20**, 2057–2063

Chu, M.-Y. and Fischer, G. A. (1962). A proposed mechanism of action of arabinofuranosylcytosine as an inhibitor of the growth of leukaemia cells. *Biochem. Pharmacol.*, **11**, 423–430

Chu, M.-Y. and Fischer, G. A. (1968). The incorporation of ^{3}H-cytosine arabinoside and its effect on murine leukaemic cells (L5178Y). *Biochem. Pharmacol.*, **17**, 753–767

Clarkson, D. R., Oppelt, W. W. and Byvoet, P. (1967). The fate of 5-iodo-2′-deoxyuridine (IUdR) in plasma and cerebrospinal fluid of dogs. *J. Pharmacol. Exp. Ther.*, **157**, 581–588

Cochran, K. W., Maassab, H. F., Tsunoda, A. and Berlin, B. S. (1965). Studies on the antiviral activity of amantadine hydrochloride. *Ann. N.Y. Acad. Sci.*, **130**, Art 1, 432–439

Coleman, V. R., Tsu, E. and Jawetz, E. (1968). "Treatment-resistance" to idoxuridine in herpetic keratitis. *Proc. Soc. Exp. Biol. Med.*, **129**, 761–765

Collins, P. and Bauer, D. J. (1977). Relative potencies of anti-herpes compounds. *Ann. N.Y. Acad. Sci.* (in press)

Conchie, A. F., Barton, B. W. and Tobin, J. O'H. (1968). Congenital cytomegalovirus infection treated with idoxuridine. *Br. Med. J.*, **4**, 162–163

Connolly, J. H., Dick, G. W. A. and Field, C. M. B. (1962). A fatal case of progressive vaccinia. *Br. Med. J.*, **1**, 1315–1317

Conomy, J. P., Beard, N. S., Matsumoto, H. and Roessmann, U. (1974). Cytarabine treatment of progressive multifocal leukoencephalopathy. Clinical course and detection of virus-like particles after antiviral chemotherapy. *J. Am. Med. Ass.*, **229**, 1313–1316

Corrigan, M. J., Gilkes, M. J. and Roberts, D.St.C. (1962). Treatment of dendritic corneal ulceration. *Br. Med. J.*, **2**, 295–298, 304–305

Corwin, M. E., Okumoto, M., Thygeson, P. and Jawetz, E. (1963). A double-blind study of the effect of 5-iodo-2'-deoxyuridine on experimental herpes simplex keratitis. *Am. J. Ophthal.*, **55**, 225–229

Craigland, J. E. (1969). Immunological response to cytomegalovirus infection in renal allograft recipients. *Am. J. Epidemiol.*, **90**, 506–513

Creasey, W. A., Papac, R. J., Markiw, M. W., Calabresi, P. and Welch, A. D. (1966). Biochemical and pharmacological studies with 1-β-D-arabino-furanosylcytosine in man. *Biochem. Pharmacol.*, **15**, 1417–1428

Dalton, A. J., Rowe, W. P., Smith, G. H., Wilsnack, R. E. and Pugh, W. E. (1968). Morphological and cytochemical studies on lymphocytic chorio-mengitis virus. *J. Virol.*, **2**, 1465–1478

Davidson, J. N. (1972). *The Biochemistry of the Nucleic Acids*, 226. (London: Chapman and Hall)

Davidson, S. I. and Evans, P. J. (1964). IDU and the treatment of herpes simplex keratitis. *Br. J. Ophthal.*, **48**, 678–683

Davies, W. L., Grunert, R. R., Haff, R. F., McGahen, J. W., Neumayer, E. M., Paulshock, M., Watts, J. C., Wood, T. R., Herrmann, E. C. and Hoffmann, C. E. (1964). Antiviral activity of 1-adamantanamine (Amantadine). *Science*, **144**, 862–863

Davies, W. L., Grunert, R. R. and Hoffmann, C. E. (1966). Influenza virus growth and antibody response in amantadine treated mice. *J. Immunol.*, **95**, 1090–1094

Davis, C. M., VanDersarl, J. V. and Coltman, C. A. (1973). Failure of cytarabine in varicella-zoster infections. *J. Am. Med. Ass.*, **224**, 122–123

Dawber, R. (1974). Idoxuridine in herpes zoster: further evaluation of inter-mittent topical therapy. *Br. Med. J.*, **1**, 526–527

Dawson, C., Togni, B. and Moore, T. E. (1968). Structural changes in chronic herpetic keratitis studied by light and electron microscopy. *Arch. Ophthal.*, **79**, 740–747

Dawson, J. R. (1933). Cellular inclusions in cerebral lesions of lethargic ence-phalitis. *Am. J. Pathol.*, **9**, 7–15

Dayan, A. D. and Lewis, P. D. (1969). Idoxuridine and jaundice. *Lancet*, **ii**, 1073

Deibel, R., Smith, R., Clarke, L. M., Decher, W. and Jacobs, J. (1974). Cytomegalovirus infections in New York State. *N.Y. State J. Med.*, **74**, 785–791

Delamore, I. W. and Prusoff, W. H. (1962). Effect of 5-iodo-2'-deoxyuridine on the biosynthesis of phosphorylated derivatives of thymidine. *Biochem. Pharmacol.*, **11**, 101–112

De Lavergne, E., Olive, D., Georges, J.-C. and Le Moyne, M.-T. (1965). Action de la 5-iodo-2'-désoxyuridine (I.D.U.) sur quelques virus à A.D.N. en cultures cellulaires. [Action of 5-iodo-2'-deoxyuridine (IDU) on certain DNA viruses in tissue culture.] *Rev. Immuno.*, **29**, 241–266

Dennis, D. T., Doberstyn, E. B., Awoke, S., Royer, G. L. and Renis, H. E. (1974). Failure of cytosine arabinoside in treating smallpox. A double-blind study. *Lancet*, **ii**, 377–374

De Rudder, J. and Privat de Garilhe, M. (1963). Inhibitory effect of some nucleosides on the growth of various human viruses in tissue culture. *Antimicrob. Ag. Chemother.*, 578–584

Dexter, D. L., Wolberg, W. H., Ansfield, F. J., Helson, L. and Heidelberger, C. (1972). The clinical pharmacology of 5-trifluoromethyl-2'-deoxyuridine. *Cancer Res.*, **32**, 247–253

Diosi, P., Moldoran, E. and Tomesen, N. (1969). Latent cytomegalovirus infection in blood donors. *Br. Med. J.*, **4**, 660–662

Ditchfield, J. and Grimyer, I. (1965). Feline rhinotracheitis virus: a feline *Herpesvirus. Virology*, **26**, 504–506

Doering, A., Keller, J. and Cohen, S. S. (1966). Some effects of D-arabinosyl nucleosides on polymer synthesis in mouse fibroblasts. *Cancer Res.*, **26**, Part 1, 2444–2450

Drach, J. C., Rentea, R. G. and Cowen, M. E. (1973). The metabolic degradation of 9-β-D-arabinofuranosyladenine (ara-A) *in vitro. Fed. Proc.*, Abstr. 3173, 777

DuBois, D. and DuBois, E. F. (1916). Clinical Calorimetry, Tenth Paper. A formula to estimate the approximate surface area if height and weight be known. *Arch. Intern. Med.*, **17**, 863–871

Duvall, C. P., Casazza, A. R., Grimley, P. M., Carbone, P. P. and Rowe, W. P. (1966). Recovery of cytomegalovirus from adults with neoplastic disease. *Ann. Intern. Med.*, **64**, 531–541

Easterbrook, K. B. (1962). Interference with the maturation of vaccinia virus by isatin β-thiosemicarbazone. *Virology*, **17**, 245–251

Eaton, M. D. and Scala, A. R. (1961). Inhibitory effect of glutamine and ammonia on replication of influenza virus in ascites tumour cells. *Virology*, **13**, 300–307

Eidinoff, M. L., Cheong, L. and Rich, M. A. (1959). Incorporation of unnatural pyrimidine bases into deoxyribonucleic acid of mammalian cells. *Science*, **129**, 1550–1551

Elliott, G. A. and Schut, A. L. (1965). Studies with cytarabine HCl (CA) in normal eyes of man, monkey and rabbit. *Am. J. Ophthal.*, **60**, 1074–1082

Engle, C. G. and Stewart, R. C. (1964). Anti-herpetic activity of 5-iodo-2'-deoxyuridine in presence of its degradation products. *Proc. Soc. Exp. Biol. Med.*, **115**, 43–45

Esiri, M. M. and Tomlinson, A. H. (1972). Herpes zoster. Demonstration of virus in trigeminal nerve and ganglion by immunofluorescence and electron microscopy. *J. Neurol. Sci.*, **15**, 35–48

Evans, A. D., Gray, O. P., Miller, M. H., Verrier Jones, E. R., Weeks, R. D. and Wells, C. E. C. (1967). Herpes simplex encephalitis treated with intravenous idoxuridine. *Br. Med. J.*, **2**, 407–410

Evans, J. S., Musser, E. A., Bostwick, L. and Mengel, G. D. (1964). The effect of 1-β-D-arabinofuranosylcytosine hydrochloride on murine neoplasms. *Cancer Res.*, **24**, 1285–1293

Evans, J. S., Musser, E. A., Mengel, G. D., Forsblad, K. R. and Hunter, J. H. (1961). Antitumour activity of 1-β-D-arabinofuranosylcytosine hydrochloride. *Proc. Soc. Exp. Biol. Med.*, **106**, 350–353

Farber, S. and Wolbach, S. B. (1932). Intranuclear and cytoplasmic inclusions ("protozoan-like bodies") in the salivary glands and other organs of infants. *Am. J. Pathol.*, **8**, 123–135

Farkas, E. (1969). Discussion of a field trial of amantadine in Hungary during an influenza A2/Hong Kong epidemic in March 1967. *Bull. Wld Hlth Org.*, **41**, 699

Farris, W. A. and Blaw, M. E. (1972). Cytarabine treatment of herpes simplex encephalitis. *Arch. Neurol.*, **27**, 99–102

Feigin, R. D., Shackelford, P. G., DeVivo, D. C. and Haymond, M. W. (1971). Floxuridine treatment of congenital cytomegalic inclusion disease. *Pediatrics*, **48**, 318–322

Feldman, L. A. and Rapp, F. (1966). Inhibition of adenovirus replication by 1-β-D-arabinofuranosylcytosine. *Proc. Soc. Exp. Biol. Med.*, **122**, 243–247

Feldman, R. A. (1969). Cytomegalovirus infection during pregnancy. A prospective study and report of six cases. *Am. J. Dis. Child.*, **117**, 517–521

Fenner, F., McAuslan, B. R., Mims, C. A., Sambrook, J. and White, D. O. (1974). *The Biology of Animal Viruses*, 207–220. (London: Academic Press, Inc.)

Feorino, P. M., Humphrey, D., Hochberg, F. and Chilicote, R. (1975). Mononucleosis-associated subacute sclerosing panencephalitis. *Lancet*, **ii**, 530–532

Ferguson, D. L. (1964). Some observations on the role of methisazone in the prophylaxis of smallpox in a rural area. *S. Afr. Med. J.*, **38**, 868–869

Fiala, M., Chow, A. W. and Guze, L. B. (1972). Susceptibility of herpesviruses to cytosine arabinoside: standardization of susceptibility test procedure and relative resistance of herpes simplex type 2 strains. *Antimicrob. Ag. Chemother.*, **1**, 354–357

Fiala, M., Chow, A. W., Miyasaki, K. and Guze, L. B. (1974). Susceptibility of herpesviruses to three nucleoside analogues and their combinations and enhancement of the antiviral effect at acid pH. *J. Infect. Dis.*, **129**, 82–85

Finkelstein, J. Z., Scher, J. and Karon, M. (1970). Pharmacologic studies of tritiated cytosine arabinoside (NSC-63878) in children. *Cancer Chemother. Rep.*, **54**, 35–39

Finklea, J. F., Hennessey, A. V. and Davenport, F. M. (1967). A field trial of amantadine prophylaxis in naturally-occurring acute respiratory illness. *Am. J. Epidemiol.*, **85**, 403–412

Fishaut, J. M., Connor, J. D. and Lampert, P. W. (1977). Comparative effects of arabinosyl nucleosides upon the postnatal growth and development of the rat. (In press)

Fishman, M. A., Haymond, M. W. and Middelkamp, J. N. (1971). Failure of idoxuridine treatment in herpes simplex encephalitis. *Am. J. Dis. Child.*, **122**, 250–252

Fletcher, R. D., Hirschfeld, J. E. and Forbes, M. (1965). A common mode of antiviral action for ammonium ions and various amines. *Nature*, **207**, 664–665

Flewett, T. H. and Ker, F. L. (1963). A case of vaccinia necrosum (or progressive vaccinia), with severe hypogammaglobulinaemia, treated with N-methylisatin beta-thiosemicarbazone (33T57). *J. Clin. Pathol.*, **16**, 271–227

Foley, J. and Williams, D. (1953). Inclusion encephalitis and its relation to subacute sclerosing leucoencephalitis: a report of five cases. *Q. J. Med.*, **22**, N.S., 157–194

Fong, C. K. Y., Hsiung, G. D. and Bensch, K. G. (1968). Productive and abortive infections of simian and nonsimian cells with a simian adenovirus SV15. II. Viral biosynthesis and effects of chemical inhibitors. *Virology*, **35**, 311–320

Foster, K. M. and Jack, I. (1969). A prospective study of the role of cytomegalovirus in post-transfusion mononucleosis. *N. Engl. J. Med.*, **280**, 1311–1316

Fujiwara, Y. and Heidelberger, C. (1970). Fluorinated pyrimidines. XXXVIII. The incorporation of 5-trifluoromethyl-2'-deoxyuridine into the deoxyribonucleic acid of vaccinia virus. *Mol. Pharmacol.*, **6**, 281–291

Fujiwara, Y., Oki, T. and Heidelberger, C. (1970). Fluorinated pyrimidines. XXXVI. Effects of 5-trifluoromethyl-2'-deoxyuridine on the synthesis of deoxyribonucleic acid of mammalian cells in culture. *Mol. Pharmacol.*, **6**, 273–280

Fulginiti, V. A., Winograd, L. A., Jackson, M. and Ellis, P. (1965). Therapy of experimental vaccinial keratitis. Effect of idoxuridine and VIG. *Arch. Ophthal.*, **74**, 539–544

Furth, J. J. and Cohen, S. S. (1968). Inhibition of mammalian DNA polymerase by the 5'-triphosphate of 1-β-D-arabinofuranosylcytosine and the 5'-triphosphate of 9-β-D-arabinofuranosyladenine. *Cancer Res.*, **28**, 2061–2067

Galbraith, A. W., Oxford, J. S., Schild, G. C., Potter, C. W. and Watson, G. I. (1971). Therapeutic effect of 1-adamantanamine HCl in naturally occurring influenza A2/Hong Kong infection. A controlled double-blind study. *Lancet*, **ii**, 113–115

Galbraith, A. W., Oxford, J. S., Schild, G. C. and Watson, G. I. (1969 a). Protective effect of 1-adamantanamine hydrochloride on influenza A2 infections in the family environment. A controlled double-blind study. *Lancet*, **ii**, 1026–1028

Galbraith, A. W., Oxford, J. S., Schild, G. C. and Watson, G. I. (1969 b). Study of 1-adamantanamine hydrochloride used prophylactically during the Hong Kong influenza epidemic in the family environment. *Bull. Wld Hlth Org.*, **41**, 677–682

Garrett, E. R., Chemburkar, P. B. and Suzuki, T. (1965). Prediction of stability in pharmaceutical preparations. XIV. The complete pH-dependent solvolytic degradations of an iodinated nucleoside, the antiviral 5-iodo-2'-deoxyuridine. *Chem. Pharm. Bull.*, **13**, 1113–1130

Garrett, E. R., Suzuki, T. and Weber, D. J. (1964). The acidic solvolytic transformation of an iodinated nucleoside, the antiviral 5-iodo-2-'-deoxyuridine. *J. Am. Chem. Soc.*, **86**, 4460–4468

Geuens, H. F. and Stephens, R. L. (1967). Influence of the pH of the urine on the rate of excretion of 1-adamantane amine. *Vth International Congress of Chemotherapy, Vienna*, June 26th–July 1st, 1967, 2/2, 703–713

Gladych, J. M. Z., Hunt, J. H., Jack, D., Haff, R. F., Boyle, J. J., Stewart, R. C. and Ferlauto, R. J. (1969). Inhibition of rhinovirus by isatin thiosemicarbazone analogues. *Nature*, **221**, 286–287

Glaser, R. and Rapp, F. (1972). Rescue of Epstein-Barr virus from somatic cell hybrids of Burkitt lymphoblastoid cells. *J. Virol.*, **10**, 288–296

Glasko, A. J. (1973). Quoted in: *Arabinosyl Nucleosides and Nucleotides*, Ch'ien, L. T., Schabel, F. M. and Alford, C. A., from W. A. Carter (ed.), *Selective Inhibitors of Viral Functions*, 227–258. (Cleveland, Ohio: Cleveland Rubber Co. Press)

Goedermans, W. T. and de Bock, C. A. (1970). Inhibition of influenza A viruses by adamantanamines: rate of resistance development in embryonated eggs and mice. *Zentbl. Bakt.*, I, **213**, 462–469

Golden, B., Bell, W. E. and McKee, A. P. (1969). Disseminated herpes simplex with encephalitis in a neonate. *J. Am. Med. Ass.*, **209**, 1219–1221

Gomez, C. P. and Sandeman, T. F. (1966). Measuring methisazone serum levels. *Lancet*, **ii**, 233

Goodman, E. L., Luby, J. P. and Johnson, M. T. (1975). Prospective double-blind evaluation of topical adenine arabinoside in male herpes progenitalis. *Antimicrob. Ag. Chemother.*, **8**, 693–697

Gordon, D. M. and Advocate, S. (1965). Vaccinial blepharokeratitis treated with cytosine arabinoside. *Am. J. Ophthal.*, **59**, 480–483

Gottschling, H. and Heidelberger, C. (1963). Fluorinated pyrimidines. XIX. Some biological effects of 5-trifluoromethyluracil and 5-trifluoromethyl-2'-deoxyuridine on *Escherichia coli* and bacteriophage T4B. *J. Mol. Biol.*, **7**, 541–560

Graham, F. L. and Whitmore, G. F. (1970). Studies in mouse L-cells on the incorporation of 1-β-D-arabinofuranosylcytosine into DNA and on inhibition of DNA polymerase by 1-β-D-arabinofuranosylcytosine 5'-triphosphate. *Cancer Res.*, **30**, 2636–2644

Greenhalgh, W. H. and Gaush, C. R. (1970). Localization of amantadine hydrochloride in tissue culture cells. *Bact. Proc., Abstr.* V107, 170

Grelak, R. P., Clark, R., Stump, J. M. and Vernier, V. G. (1970). Amantadine-dopamine interaction: possible mode of action in parkinsonism. *Science*, **169**, 203–204

Griswold, D. E., Heppner, G. H. and Calabresi, P. (1972). Selective suppression of humoral and cellular immunity with cytosine arabinoside. *Cancer Res.*, **32**, 298–301

Grossgebauer, K. and Langmaack, H. (1970). Failure of 1-adamantanamine (Symmetrel) to modify influenza virus-induced pyrogenicity. *Arch. Ges. Virusforsch.*, **31**, 385–386

Grunert, R. R., McGahen, J. W. and Davies, W. L. (1965). The *in vivo* antiviral activity of 1-adamantanamine (Amantadine). I. Prophylactic and therapeutic activity against influenza viruses. *Virology*, **26**, 262–269

Gudnadóttir, M., Helgadóttir, H., Bjarnason, O. and Jónsdóttir, K. (1964). Virus isolated from the brain of a patient with multiple sclerosis. *Exp. Neurol.*, **9**, 85–95

Haas, R. and Maass, G. (1964). Die Wirkung von 5-Iod-2'-desoxyuridin auf die Vermehrung von SV-40 in Gewebekulturen. *Arch. Ges. Virusforsch.*, **14**, 567–582

Haase, A. T. and Levinson, W. (1973). Inhibition of RNA slow viruses by thiosemicarbazones. *Biochem. Biophys. Res. Commun.*, **51**, 875–880

Hall, T. C., Wilfert, C., Jaffe, N., Traggis, D., Lux, S., Rompf, P. and Katz, S. (1969). Treatment of varicella-zoster with cytosine arabinoside. *Trans. Ass. Am. Phys.*, **82**, 201–210

Hall-Smith, S. P., Corrigan, M. J. and Gilkes, M. J. (1962). Treatment of herpes simplex with 5-iodo-2′-deoxyuridine. *Br. Med. J.*, **2**, 1515–1516

Handschumacher, R. E., Calabresi, P., Welch, A. D., Bono, V., Fallon, H. and Frei, E. (1962). Summary of current information on 6-azauridine. *Cancer Chemother. Rep.*, **21**, 1–11

Hanka, L. J., Kuentzel, S. L. and Neil, G. L. (1970). Improved microbiological assay for cytosine arabinoside (NSC-63878). *Cancer Chemother Rep.*, **54**, 393–397

Hansson, O., Johansson, S. G. O. and Vahlquist, B. (1966). Vaccinia gangrenosa with normal humoral antibodies. A case possibly due to deficient cellular immunity treated with N-methylisatin β-thiosemicarbazone (compound 33T57, Marboran). *Acta Paediatr. Scand.*, **55**, 264–272

Haslam, R. H. A., McQuillen, M. P. and Clark, D. B. (1969). Amantadine therapy in subacute sclerosing panencephalitis. *Neurology*, **19**, 1080–1086

Hedley-White, E. T., Smith, B. P., Tyler, H. R. and Peterson, W. P. (1966). Multifocal leukoencephalopathy with remission and five year survival. *J. Neuropath. Exp. Neurol.*, **25**, 107–116

Heidelberger, C. and Anderson, S. W. (1964). Fluorinated pyrimidines. XXI. The tumor-inhibitory activity of 5-trifluoromethyl-2′-deoxyuridine. *Cancer Res.*, **24**, 1979–1985

Heidelberger, C., Birnie, G. D., Boohar, J. and Wentland, D. (1963). Fluorinated pyrimidines. XX. Inhibition of the nucleoside phosphorylase cleavage of 5-fluoro-2′-deoxyuridine by 5-trifluoromethyl-2′-deoxyuridine. *Biochim. Biophys. Acta*, **76**, 315–318

Heidelberger, C., Boohar, J. and Birnie, G. (1964 a). Fluorinated pyrimidines. XXII. Effects of various compounds on the incorporation of [C^{14}] formate into DNA thymine in suspensions of Ehrlich ascites cells. *Biochim. Biophys. Acta*, **91**, 636–638

Heidelberger, C., Boohar, J. and Kampschroer, B. (1965). Fluorinated pyrimidines. XXIV. *In vivo* metabolism of 5-trifluoromethyluracil-2-C^{14} and 5-trifluoromethyl-2′-deoxyuridine-2-C^{14}. *Cancer Res.*, **25**, 377–381

Heidelberger, C., Parsons, D. G. and Remy, D. C. (1964 b). Syntheses of 5-trifluoromethyluracil and 5-trifluoromethyl-2′-deoxyuridine. *J. Med. Chem.*, **7**, 1–5

Heiner, G. G., Fatima, N., Russell, P. K., Haase, A. T., Ahmad, N., Mohammed, N., Thomas, D. B., Mack, T. M., Khan, M. M., Knatterud, G. L., Anthony, R. L. and McCrumb, F. R. (1971). Field trials of methisazone as a prophylactic agent against smallpox. *Am. J. Epidemiol.*, **94**, 435–449

Helson, L., Yagoda, A., McCarthy, M., Murphy, M. L. and Krakoff, I. H. (1970). Clinical trials with trifluoromethyl-2′-deoxyuridine (F$_3$TdR). *Proc. Am. Ass. Cancer Res.*, **10**, 35

Hermodson, J. and Dinter, Z. (1962). Properties of bovine virus diarrhoea virus. *Nature*, **194**, 893–894

Herrmann, E. C. (1961). Plaque inhibition test for detection of specific inhibitors of DNA containing viruses. *Proc. Soc. Exp. Biol. Med.*, **107**, 142–145

Herrmann, E. C. (1968). Sensitivity of herpes simplex virus, vaccinia virus and adenoviruses to deoxyribonucleic acid inhibitors and thiosemicarbazones in a plaque suppression test. *Appl. Microbiol.*, **16**, 1151–1155

Herrmann, E. C., Gabliks, J., Engle, C. and Perlman, P. L. (1960). Agar diffusion method for detection and bioassay of antiviral antibiotics. *Proc. Soc. Exp. Biol. Med.*, **103**, 625–628

Hildebrandt, R. J., Sever, J. L., Margileth, A. M. and Callaghan, D. A. (1967). Cytomegalovirus in the normal pregnant woman. *Am. J. Obstet. Gynecol.*, **98**, 1125–1128

Hill, R. B., Rowlands, D. T. and Rifkind, D. (1964). Infectious pulmonary disease in patients receiving immunosuppressive therapy for organ transplantation. *N. Engl. J. Med.*, **271**, 1021–1027

Hirschman, S. Z. (1969). Effect of cytosine arabinoside on the replication of the Moloney sarcoma virus in 3T3 cell cultures. *Proc. Am. Ass. Cancer Res.*, **10**, 38

Ho, D. H. W. and Frei, E. (1971). Clinical pharmacology of 1-β-D-arabinofuranosyl cytosine. *Clin. Pharmacol. Ther.*, **12**, 944–954

Hoffmann, C. E., Neumayer, E. M., Haff, R. F. and Goldsby, R. A. (1965). Mode of action of the antiviral activity of amantadine in tissue culture. *J. Bact.*, **90**, 623–628

Hornick, R. B., Togo, Y., Mahler, S. and Iezzoni, D. (1970). Evaluation of amantadine hydrochloride in the treatment of A2 influenzal disease. *Ann. N.Y. Acad. Sci.*, **173**, Art. 1, 10–19

Horta-Barbosa, L., Fuccillo, D. A., Sever, J. L. and Zeman, W. (1969). Subacute sclerosing panencephalitis: isolation of measles virus from a brain biopsy. *Nature*, **221**, 974–975

Hossain, M. S., Hryniuk, W., Foerster, J., Israels, L. G., Chowdhury, A. S. and Biswas, M. K. (1972). Treatment of smallpox with cytosine arabinoside. *Lancet*, **ii**, 1230–1232

Hryniuk, W., Foerster, J., Shojania, M. and Chow, A. (1972). Cytarabine for herpesvirus infections. *J. Am. Med. Ass.*, **219**, 715–718

Huberman, E. and Heidelberger, C. (1972). The mutagenicity to mammalian cells of pyrimidine nucleoside analogs. *Mutation Res.*, **14**, 130–132

Hubert-Habart, M. and Cohen, S. S. (1962). The toxicity of 9-β-D-arabinofuranosyladenine to purine-requiring *Escherichia coli*. *Biochim. Biophys. Acta*, **59**, 468–471

Hyndiuk, R. A., Kaufman, H. E., Ellison, E. and Centifanto, Y. (1968). Newer antiviral agents—comparative assays *in vivo* and *in vitro*. *Chemotherapy*, **13**, 139–145

Ilan, J., Tokuyasu, K. and Ilan, J. (1970). Phosphorylation of D-arabinosyl adenine by *Plasmodium berghei* and its partial protection of mice against malaria. *Nature*, **228**, 1300–1301

Illis, L. S. and Gostling, J. V. T. (1972). *Herpes simplex encephalitis*. (Bristol: Scientechnica)

Illis, L. S. and Merry, R. T. G. (1972). Treatment of herpes simplex encephalitis. *J. R. Coll. Phys. Lond.*, **7**, 34–44

Inagaki, A., Nakamura, T. and Wakisaka, G. (1969). Studies on the mechanism of action of 1-β-D-arabinofuranosylcytosine as an inhibitor of DNA synthesis in human leukocytes. *Cancer Res.*, **29**, 2169–2176

Inoue, Y. K. (1975). An avian-related new herpesvirus infection in man—subacute myelo-optico-neuropathy (SMON). In: J. L. Melnick and F. Rapp (eds.) *Progress in Medical Virology*, 35–42 (London: S. Karger)

Jack, M. K. and Sorenson, R. W. (1963). Vaccinial keratitis treated with IDU. *Arch. Ophthal.*, **69**, 730–732

Jackson, G. G., Muldoon, R. L. and Akers, L. W. (1964). Serological evidence for prevention of influenzal infection in volunteers by an anti-influenzal drug adamantanamine hydrochloride. *Antimicrob. Ag. Chemother.*, 703–707

Jackson, G. G., Stanley, E. D. and Muldoon, R. L. (1967). Chemoprophylaxis of viral respiratory diseases. *First International Conference on Vaccines against Viral and Rickettsial Diseases of Man.* PAHO/WHO Scientific Publication No. 147, 595–603

Jackson, N. (1963). Treatment of herpes simplex of the skin with 5-iodo-2′-deoxyuridine ("Kerecid"). *J. Irish Med. Ass.*, **52**, 156–157

Janković, T., Šuvaković, V., Kecmanović, M., Džibo, D., Lazarević, P., Ristić, S., Ranitović, S., Banković, A. and Radovanović, Z. (1972). Iskustva s profilaksom variole Marboranom i hiperimunim gama-globulinom 1972. Godine. In: L. Stojković (ed.), *Variola u Jugoslaviji 1972. godine*, 275–279. [Experience in the prophylaxis of variola with Marboran and hyperimmune gamma-globulin in 1972.]
(Ljubljana: ČGP Delo)

Jaroszyńska-Weinberger, B. (1970). Treatment with methisazone of complications following smallpox vaccination. *Arch. Dis. Child.*, **45**, 573–580

Jaroszyńska-Weinberger, B. and Mészáros, J. (1966). A comparison of the protective effect of methisazone and a hyperimmune antivaccinial gamma-globulin in primary smallpox vaccination carried out in the presence of contraindications. *Lancet*, **i**, 948–951

Jensen, E. M., Force, E. F. and Unger, J. B. (1961). Inhibitory effect of ammonium ions on influenza virus in tissue cultures. *Proc. Soc. Exp. Biol. Med.*, **107**, 447–451

Jensen, E. M. and Liu, O. C. (1961). Studies of inhibitory effect of ammonium ions in several virus-tissue culture sytems. *Proc. Soc. Exp. Biol. Med.*, **107**, 834–838

Jensen, E. M. and Liu, O. C. (1963). Inhibitory effect of simple aliphatic amines on influenza virus in tissue culture. *Proc. Soc. Exp. Biol. Med.*, **112**, 456–459

Jepson, C. N. (1964). Treatment of herpes simplex of the cornea with IDU. A double-blind study. *Am. J. Ophthal.*, **57**, 213–217

John, R. W. and Tobin, J. O'H. (1973). Neonatal *Herpesvirus hominis* infections. *Postgrad. Med. J.*, **49**, 380–382

Johnson, M. T., Luby, J. P., Buchanan, R. A. and Mikulec, D. (1975). Treatment of varicella-zoster virus infections with adenine arabinoside. *J. Infect. Dis.*, **131**, 225–229

Joncas, J. H., Menezes, J. and Huang, E. S. (1975). Persistence of CMV genome in lymphoid cells after congenital infections. *Nature*, **258**, 432–433

Jones, B. R. and Al-Hussaini, M. Ķ. (1963). Therapeutic considerations in ocular vaccinia. *Trans. Ophthal. Soc.*, **83**, 613–631

Juel-Jensen, B. E. (1970). The natural history of shingles. *J. R. Coll. Gen. Practit.*, **20**, 323–327

Juel-Jensen, B. E. (1971). Herpetic whitlows: a medical risk. *Br. Med. J.*, **2**, 681

Juel-Jensen, B. E. and MacCallum, F. O. (1964). Treatment of herpes simplex lesions of the face with idoxuridine: results of a double-blind controlled trial. *Br. Med. J.*, **2**, 987–988

Juel-Jensen, B. E. and MacCallum, F. O. (1972). *Herpes simplex, Varicella and Zoster.* (London: William Heinemann Medical Books, Ltd)

Juel-Jensen, B. E., MacCallum, F. O., Mackenzie, A. M. and Pike, M. C. (1970). Treatment of zoster with idoxuridine in dimethyl sulphoxide. Results of two double-blind controlled trials. *Br. Med. J.*, **2**, 776–780

Jurna, I., Grossmann, W. and Nell, T. (1972). Depression by amantadine of drug-induced rigidity in the rat. *Neuropharmacology*, **11**, 559–564

Kääriäinen, L., Klemola, E. and Paloheimo, J. (1966). Rise of cytomegalovirus antibodies in an infectious-mononucleosis-like syndrome after transfusion. *Br. Med. J.*, **1**, 1270–1272

Kaji, M., Yanaga, T., Ito, M., Arita, M., Saeki, K., Tanoue, M. and Mashiba, H. (1966). Effects of 1-adamantanamine on the electrocardiogram and trans-membrane potential of the rabbit heart. *Fukuoka Acta Med.*, **57**, 251–258

Kaplan, A. S. and Ben-Porat, T. (1966). Mode of action of 5-iodouracil deoxyriboside. *J. Mol. Biol.*, **19**, 320–322

Karnofsky, D. A. and Lacon, D. R. (1966). The effects of 1-β-D-arabinofurano-sylcytosine on the developing chick embryo. *Biochem. Pharmacol.*, **15**, 1435–1442

Kato, N. and Eggers, H. J. (1969). Inhibition of uncoating of fowl plague virus by 1-adamantanamine hydrochloride. *Virology*, **37**, 632–641.

Katz, E., Margalith, E., Winer, B. and Goldblum, N. (1973 a). Synthesis of vaccinia virus polypeptides in the presence of isatin β-thiosemicarbazone. *Antimicrob. Ag. Chemother.*, **4**, 44–48

Katz, E., Winer, B., Margalith, E. and Goldblum, N. (1973 b). Isolation and characterization of an IBT-dependent mutant of vaccinia virus. *J. Gen. Virol.*, **19**, 161–164

Kaufman, H. E. (1962). Clinical cure of herpes simplex keratitis by 5-iodo-2′-deoxyuridine. *Proc. Soc. Exp. Biol. Med.*, **109**, 251–252

Kaufman, H. E., Brown, D. C. and Ellison, E. D. (1967). Recurrent herpes in the rabbit and man. *Science*, **156**, 1628–1629

Kaufman, H. E., Brown, D. C. and Ellison, E. D. (1968). Herpes virus in the lacrimal gland, conjunctiva and cornea of man—a chronic infection. *Am. J. Ophthal.*, **65**, 32–35

Kaufman, H. E., Capella, J. A., Maloney, E. D., Robbins, J. E., Cooper, G. M. and Uotila, M. H. (1964). Corneal toxicity of cytosine arabinoside. *Arch. Ophthal.*, **72**, 535–540

Kaufman, H. E. and Heidelberger, C. (1964). Therapeutic antiviral action of 5-trifluoromethyl-2′-deoxyuridine in herpes simplex keratitis. *Science*, **145**, 585–586

Kaufman, H. E. and Maloney, E. D. (1963 a). IDU and cytosine arabinoside in experimental herpetic keratitis. *Arch. Ophthal.*, **69**, 126–129

Kaufman, H. E. and Maloney, E. D. (1963 b). Therapeutic antiviral activity in tissue culture. *Proc. Soc. Exp. Biol. Med.*, **112**, 4–7

Kaufman, H. E., Martola, E.-L. and Dohlman, C. H. (1963). Herpes simplex treatment with IDU and corticosteroids. *Arch. Ophthal.*, **69**, 468–472

Kaufman, H. E., Nesburn, A. B. and Maloney, E. D. (1962 a). IDU therapy of herpes simplex. *Arch. Ophthal.*, **67**, 583–591

Kaufman, H. E., Nesburn, A. B. and Maloney, E. D. (1962 b). Cure of vaccinia infection by 5-iodo-2'-deoxyuridine. *Virology*, **18**, 567–569

Keeble, S. J., Christofinis, G. J. and Wood, W. (1958). Natural virus-B infection in monkeys. *J. Pathol. Bact.*, **76**, 189–199

Kempe, C. H. (1960). Studies on smallpox and complications of smallpox vaccination. *Pediatrics*, **26**, 176–189.

Kempe, C. H., Fulginiti, V. and Sieber, O. (1967). Chemotherapy of life-threatening dermal complications of smallpox vaccination. *Vth International Congress of Chemotherapy, Vienna*, Abstracta Pt. 2, 1234

Kempe, C. H., Rodgerson, D. and Sieber, O. F. (1965). Measurement of *N*-methylisatin β-thiosemicarbazone serum levels. *Lancet*, **i**, 824

Kihlman, B. A., Nichols, W. W. and Levan, A. (1963). The effect of deoxyadenosine and cytosine arabinoside on the chromosomes of human leukocytes in vitro. *Hereditas*, **50**, 139–143

Kim, J. H. and Eidinoff, M. L. (1965). Action of 1-β-D-arabinofuranosylcytosine on the nucleic acid metabolism and viability of HeLa cells. *Cancer Res.*, **25**, 698–702

Kimball, A. P., Bowman, B., Bush, P. S., Herriot, J. and LePage, G. A. (1966). Inhibitory effects of the arabinosides of 6-mercaptopurine and cytosine on purine and pyrimidine metabolism. *Cancer Res.*, **26**, 1337–1343

Kimball, A. P. and Wilson, M. J. (1968). Inhibition of DNA polymerase by β-D-arabinosylcytosine and reversal of inhibition by deoxycytidine-5'-triphosphate. *Proc. Soc. Exp. Biol. Med.*, **127**, 429–432

Kitamoto, O. (1971). Therapeutic effectiveness of amantadine hydrochloride in naturally occurring Hong Kong influenza—double-blind studies. *Jap. J. Tuberc.*, **17**, 1–7

Kitamoto, O., Hirayama, M., Ichida, F., Kanamitsu, M., Ishida, N., Kaji, M., Kawana, R., Kitta, A., Kono, R., Matsumoto, Y., Nakamura, K., Nakamura, T., Okuda, R., Shigematsu, I., Sonoguchi, T., Subiura, A., Tokuda, M. and Uchida, T. (1970). Therapeutic effectiveness of amantadine HCl in influenza A2 and influenza A Hong Kong double-blind studies. *Proc. 6th Int. Congr. Chemother.*, **2**, 65–70

Knight, V., Fedson, D., Baldini, J., Douglas, R. B. and Couch, R. B. (1970). Amantadine therapy of epidemic influenza A2/Hong Kong. *Antimicrob. Ag. Chemother.*, 370–371

Knotts, F. B., Cook, M. L. and Stevens, J. G. (1973). Latent herpes simplex virus in the central nervous system of rabbits and mice. *J. Exp. Med.*, **138**, 740–744

Kolb, K. E., Bower, R. K. and Duffy, C. E. (1963). Effect of 5-iodo-2'-deoxyuridine on pseudorabies infection in rabbits. *Proc. Soc. Exp. Biol. Med.*, **113**, 476–478

Koplan, J. P., Monsur, K. A., Foster, S. O., Huq, F., Rahaman, M. M., Huq, S., Buchanan, R. A. and Ward, N. A. (1975). Treatment of variola major with adenine arabinosine. *J. Infect. Dis.*, **131**, 34–39

Kurtz, S. M., Fisken, R. A., Kaump, D. H. and Schardein, J. L. (1968). Toxicity of 9-β-D-arabinofuranosyladenine in mice and rabbits. *Antimicrob. Ag. Chemother.*, 180–189

Kury, G. and Crosby, R. J. (1967). The teratogenic effect of 5-trifluoromethyl-2'-deoxyuridine in chicken embryos. *Toxicol. Appl. Pharmacol.*, **11**, 72–80

Ladniy, I. D. (1974). [The use of methisazone for the prophylaxis of post-vaccinial complications following smallpox vaccination. In Russian.] *Zh. Mikrobiol. Epidemiol. Immunobiol.*, No. 1, 46–49

Lagerkvist, B. and Ekelund, H. (1975). Cytarabine treatment of herpes simplex encephalitis in infants and small children. A report on three cases with a short review of the literature. *Scand. J. Infect. Dis.*, **7**, 81–84

Ledinko, N. (1967). Plaque assay of the effects of cytosine arabinoside and 5-iodo-2'-deoxyuridine on the synthesis of H-1 virus particles. *Nature*, **214**, 1346–1347

Lee, W. W., Benitez, A., Goodman, L. and Baker, B. R. (1960). Potential anticancer agents. XL. Synthesis of the β-anomer of 9-(D-arabinofuranosyl) adenine. *J. Am. Chem. Soc.*, **82**, 2648–2649

Leonard, L. L., ter Meulen, V. and Freeman, J. M. (1971). *In vitro* sensitivity of measles virus to 6-azauridine. *Proc. Soc. Exp. Biol. Med.*, **136**, 857–862

LePage, G. A. (1970). Arabinosyladenine and arabinosylhypoxanthine metabolism in murine tumor cells. *Can. J. Biochem.*, **48**, 75–78.

LePage, G. A. and Junga, I. G. (1965). Metabolism of purine nucleoside analogs. *Cancer Res.*, **25**, Pt. 1, 46–52

LePage, G. A., Khaliq, A. and Gottlieb, J. A. (1973). Studies of 9-β-D-arabino-furanosyladenine in man. *Drug Metab. Disposition*, **1**, 756–759

Lerner, A. M. and Bailey, E. J. (1972). Concentrations of idoxuridine in serum, urine, and cerebrospinal fluid of patients with suspected diagnoses of *Herpesvirus hominis* encephalitis. *J. Clin. Invest.*, **51**, 45–49

Levinsohn, E. M., Foy, H. M., Kenny, G. E., Wentworth, B. B. and Grayston, J. T. (1969). Isolation of cytomegalovirus from a cohort of 100 infants throughout the first year of life. *Proc. Soc. Exp. Biol. Med.*, **132**, 957–962

Levinson, W., Faras, A., Morris, R., Mikelens, P., Ringold, G., Kass, S., Levinson, B. and Jackson, J. (1973 a). Effect of N-methyl isatin β-thiosemi-carbazone on Rous sarcoma virus, several isolated enzymes and other viruses. In: C. F. Fox and W. S. Robinson (eds.) *Virus Research. Second ICN-UCLA Symposium on Molecular Biology*, 403–413. (New York and London: Academic Press)

Levinson, W., Faras, A., Woodson, B., Jackson, J. and Bishop, J. M. (1973 b). Inhibition of RNA-dependent DNA polymerase of Rous sarcoma virus by thiosemicarbazones and several cations. *Proc. Natnl. Acad. Sci.*, **70**, 164–168

Lieberman, M. W., Verbin, R. S., Landay, M., Liang, H., Farber, E. and Lee, T. N. (1970). A probable role for protein synthesis in intestinal epithelial cell damage induced *in vivo* by cytosine arabinoside, nitrogen mustard, or X-irradiation. *Cancer Res.*, **30**, 942–951

Likar, M. (1970). Effectiveness of amantadine in protecting vaccinated volunteers from an attenuated strain of influenza A2/Hong Kong virus. *Ann. N.Y. Acad. Sci.*, **173**, Art. 1, 108–112

Lindberg, B., Klenow, H. and Hansen, K. (1967). Some properties of partially purified mammalian adenosine kinase. *J. Biol. Chem.*, **242**, 350–356

Little, J. M., Lorenzetti, D. W. C., Brown, D. C., Schweem, H. H., Jones, B. R. and Kaufman, H. E. (1968). Studies of adenovirus type III infection treated with methisazone and trifluorothymidine. *Proc. Soc. Exp. Biol. Med.*, **127**, 1028–1032

Loddo, B., Schivo, M. L. and Ferrari, W. (1963). Development of vaccinia virus resistant to 5-iodo-2'-deoxyuridine. *Lancet*, **ii**, 914–915

Logan, J. C., Fox, M. P., Morgan, J. H., Makohon, A. M. and Pfau, C. J. (1975). Arenavirus inactivation on contact with N-substituted isatin beta-thiosemicarbazones and certain cations. *J. Gen. Virol.*, **28**, 271–283

Long, W. F. and Burke, D. C. (1969). The effect of infection with fowl plague virus on protein synthesis in chick embryo cells. *J. Gen. Virol.*, **6**, 1–14

Long, W. F. and Olusanya, J. (1972). Adamantamine and early events following influenza virus infection. *Arch. Ges. Virusforsch.*, **36**, 18–22

Lopez, C., Simmons, R. L., Mauer, S. M., Najarian, J. S. and Good, R. H. (1974). Association of renal allograft rejection with virus infections. *Am. J. Med.*, **56**, 280–289

Lowry, D. R., Rowe, W. P., Teich, N. and Hartley, J. W. (1971). Murine leukaemia virus: high-frequency activation in vitro by 5-iododeoxyuridine and 5-bromodeoxyuridine. *Science*, **174**, 155–156

Lowry, S. P., Melnick, J. L. and Rawls, W. E. (1971). Investigation of plaque formation in chick embryo cells as a biological marker for distinguishing herpes virus type 2 from type 1. *J. Gen. Virol.*, **10**, 1–9

Luby, J. P., Johnson, M. T., Buchanan, R., Ch'ien, L. T., Whitley, R. and Alford, C. (1975). Adenine arabinoside therapy of varicella-zoster virus infections. Summary of phase II studies. In: D. Pavan-Langston, R. A. Buchanan and C. A. Alford (eds.) *Adenine Arabinoside; an Antiviral Agent*, 237–245. (New York: Raven Press)

Lwoff, A. and Lwoff, M. (1964). Remarques sur l'isatine β-thiosemicarbazone et sur quelques inhibiteurs du developpement viral. *C.R. Hebd. Séanc. Acad. Sci., Paris*, **258**, 1924–1927

Maassab, H. F. and Cochran, K. W. (1964). Rubella virus: inhibition in vitro by amantadine hydrochloride. *Science*, **145**, 1443–1444

MacCallum, F. O. and Juel-Jensen, B. E. (1966). Herpes simplex virus skin infection in man treated with idoxuridine in dimethyl sulphoxide. Results of double-blind controlled trial. *Br. Med. J.*, **2**, 805–807

Magee, W. E. and Bach, M. K. (1965). Biochemical studies on the antiviral activities of the isatin β-thiosemicarbazones. *Ann. N.Y. Acad. Sci.*, **130**, Art. 1, 80–91

Mainwaring, D. (1962). Eczema vaccinatum. *Br. Med. J.*, **1**, 1412–1413

Manilla, G. T., Hood, T. K. and Eakin, N. R. (1965). Treatment of plantar warts. *Rocky Mountain Med. J.*, **62**, 42

Mann, J. R. (1971). Cytosine arabinoside and herpes zoster. *Lancet*, **ii**, 166

Marsh, W. B. and Mitchell, L. (1965). Generalized vaccinia and N-methylisatin β-thiosemicarbazone. *Med. J. Aust.*, **1**, 947–948

Marshall, W. J. S. (1967). Herpes simplex encephalitis treated with idoxuridine and external decompression. *Lancet*, **ii**, 579–580

Mastan, P. F. and Henderson, J. W. (1966). Penetration of idoxuridine into the anterior segment after transdermal subconjunctival injection. *Invest. Opthal.*, **5**, 320–321

Máté, J., Simon, M. and Juvancz, I. (1971). Use of Viregyt (amantadine hydrochloride) in the treatment of epidemic influenza. *Ther. Hung.*, **19**, 117–121

Máté, J., Simon, M., Juvancz, I., Takátsy, G., Hollós, I. and Farkas, E. (1970). Prophylactic use of amantadine during Hong Kong influenza epidemic. *Acta Microbiol. Acad. Sci. Hung.*, **17**, 285–296

Mathias, A. P., Fischer, G. A. and Prusoff, W. H. (1959). Inhibition of the growth of mouse leukaemia cells in culture by 5-iododeoxyuridine. *Biochim. Biophys. Acta*, **36**, 560–561

Maxwell, E. (1963). Treatment of corneal disease with 5-iodo-2'-deoxyuridine (IDU). A clinical evaluation of 500 cases. *Am. J. Ophthal.*, **55**, 237–238

McCracken, G. H. and Luby, J. P. (1972). Cytosine arabinoside in the treatment of congenital cytomegalic inclusion disease. *J. Pediatr.*, **80**, 488–495

McCracken, G. H., Shinefield, H. R., Cobb, K., Rausen, A. R., Dische, M. R. and Eichenwald, H. F. (1969). Congenital cytomegalic inclusion disease. A longitudinal study of 20 patients. *Am. J. Dis. Child.*, **117**, 522–539

McGill, J. I. (1975). The treatment of herpetic corneal ulceration. *Update*, 531–540

McGill, J., Holt-Wilson, A. D., McKinnon, J. R., Williams, H. P. and Jones, B. R. (1974 a). Some aspects of the clinical use of trifluorothymidine in the treatment of herpetic ulceration of the cornea. *Trans. Ophthal. Soc. UK*, **94**, 342–352

McGill, J., Williams, J., McKinnon, J., Holt-Wilson, A. D. and Jones, B. R. (1974 b). Reassessment of idoxuridine therapy of herpetic keratitis. *Trans. Ophthal. Soc. UK*, **94**, 542–552

McKelvey, E. M. and Kwaan, H. C. (1969). Cytosine arabinoside therapy for disseminated herpes zoster in a patient with IgG pyroglobulinemia. *Blood*, **34**, 706–711

McLaren, C. and Potter, C. W. (1973). In vitro inhibition of influenza virus A/Mill Hill/1/72 by amantadine. *Lancet*, **i**, 1157

McMath, W. F. T. and Wilson, H. T. H. (1965). Cowpox treated with Marboran (methisazone). *Br. Med. J.*, **1**, 1041–1042

McNeill, T. A. (1972). Effect of methisazone and other drugs on mouse hemopoietic colony formation in vitro. *Antimicrob. Ag. Chemother.*, **1**, 6–11

McNeill, T. A. (1973). Inhibition of granulocyte-macrophage colony formation in vitro by substituted isatin thiosemicarbazones. *Antimicrob. Ag. Chemother.*, **4**, 105–108

McNeill, T. A., Fleming, W. A., McClure, S. F. and Killen, M. (1972). Suppression of immune and hemopoietic cellular responses by methisazone. *Antimicrob. Ag. Chemother.*, **1**, 1–5

Melnick, J. L. (1974). Classification and nomenclature of viruses, 1974. *Progr. Med. Virol.*, **17**, 290–294

Metz, H., Gregoriou, H. and Sandifer, P. (1964). Subacute sclerosing panencephalitis: a review of 17 cases with special reference to clinical diagnostic criteria. *Arch. Dis. Child.*, **39**, 554–557

Miller, F. A. (1967). Inhibition of B virus in cell culture by 5-iodo-2'-deoxyuridine. *Appl. Microbiol.*, **15**, 733–735

Miller, F. A., Dixon, G. J., Ehrlich, J., Sloan, B. J. and McLean, I. W. (1969). Antiviral activity of 9-*β*-D-arabinofuranosyladenine. I. Cell culture studies. *Antimicrob. Ag. Chemother.*, 136–147

Minocha, H. C. and Maloney, B. (1970). Inhibition of fibroma viral deoxyribonucleic acid synthesis by fluorodeoxyuridine and cytosine arabinoside. *Am. J. Vet. Res.*, **31**, 1469–1475

Minton, S. A., Officer, J. E. and Thompson, R. L. (1953). Effect of thiosemicarbazones and dichlorophenoxy thiouracil on multiplication of a recently isolated strain of variola-vaccinia virus in the brain of the mouse. *J. Immunol.*, **70**, 222–228

Mitchell, M. S., Kaplan, S. R. and Calabresi, P. (1969 a). Alteration of antibody synthesis in the rat by cytosine arabinoside. *Cancer Res.*, **29**, 896–904

Mitchell, M. S., Wade, M. E., DeConti, B. C., Bertino, J. R. and Calabresi, P. (1969 b). Immunosuppressive effects of cytosine arabinoside and methotrexate in man. *Ann. Intern. Med.*, **70**, 535–547

Momparler, R. L. (1969). Effect of cytosine arabinoside 5'-triphosphate on mammalian DNA polymerase. *Biochem. Biophys. Res. Commun.*, **34**, 464–471

Momparler, R. L. and Fischer, G. A. (1968). Mammalian deoxynucleoside kinases. I. Deoxycytidine kinase: purification, properties and kinetic studies with cytosine arabinoside. *J. Biol. Chem.*, **243**, 4298–4304

Montgomery, R., Youngblood, L. and Medearis, D. N. (1972). Recovery of cytomegalovirus from the cervix in pregnancy. *Pediatrics*, **49**, 524–531

Moore, E. C. and Cohen, S. S. (1967). Effects of arabinonucleotides on ribonucleotide reduction by an enzyme system from rat tumor. *J. Biol. Chem.*, **242**, 2116–2118

Morris, N. R. and Cramer, J. W. (1966). DNA synthesis by mammalian cells inhibited in culture by 5-iodo-2'-deoxyuridine. *Mol. Pharmacol.*, **2**, 1–9

Munro, T. W. and Sabina, L. R. (1970). Inhibition of infectious bovine rhinotracheitis virus multiplication by thiosemicarbazones. *J. Gen. Virol.*, **7**, 55–63

Munyon, W., Hughes, R., Angermann, J., Bereczky, E. and Dmochowski, L. (1964). Studies on the effect of 5-iododeoxyuridine and *p*-fluorophenylalanine on polyoma-virus formation *in vitro*. *Cancer Res.*, **24**, 1880–1886

Munyon, W., Kraiselburd, E., Davis, D. and Mann, J. (1971). Transfer of thymidine kinase to thymidine kinaseless L cells by infection with ultraviolet-irradiated herpes simplex virus. *J. Virol.*, **7**, 813–820

Nafta, I., Turcanu, A. G., Braun, I., Companetz, W., Simionescu, A., Birt, E. and Florea, V. (1970). Administration of amantadine for the prevention of Hong Kong influenza. *Bull. Wld. Hlth. Org.*, **42**, 423–427

Nagington, J. (1971). Cytomegalovirus antibody production in renal transplant patients. *J. Hyg., Camb.*, **69**, 645–660

Nahmias, A. J., Alford, C. A. and Korones, S. B. (1970). Infection of the newborn with herpesvirus hominis. *Adv. Pediatr.*, **17**, 185–226

Najjar, T. A., Sleeper, H. R. and Calabresi, P. (1969). The use of 5-iodo-2′-deoxyuridine (IUDR) in Orabase and Plastibase for treatment of oral herpes simplex. *J. Oral Med.*, **24**, 53–57

Nestler, H. J. and Garrett, E. R. (1968). Prediction of stability in pharmaceutical preparations. XV. Kinetics of hydrolysis of 5-trifluoromethyl-2′-deoxyuridine. *J. pharm. Sci.*, **57**, 1117–1125

Neumayer, E. M., Haff, R. F. and Hoffmann, C. E. (1965). Antiviral activity of amantadine hydrochloride in tissue culture and *in ovo*. *Proc. Soc. Exp. Biol. Med.*, **119**, 393–396

Nichols, W. W. (1964). *In vitro* chromosome breakage induced by arabinosyl-adenine in human leukocytes. *Cancer Res.*, **24**, Pt. 1, 1502–1504

Nolan, D. C., Lauter, C. B. and Lerner, A. M. (1973). Idoxuridine in herpes simplex virus (type 1) encephalitis. Experience with 29 cases in Michigan, 1966 to 1971. *Ann. Intern. Med.*, **78**, 243–246

Notari, R. E. (1967). A mechanism for the hydrolytic deamination of cytosine arabinoside in aqueous buffer. *J. Pharm. Sci.*, **56**, 804–809

Numazaki, Y., Yano, N., Morizuka, T., Takai, S. and Ishida, N. (1970). Primary infection with human cytomegalovirus: virus isolation from healthy infants and pregnant women. *Am. J. Epidemiol.*, **91**, 410–417

O'Donoghue, J. M., Ray, C. G., Terry, D. and Beaty, H. M. (1972). Prevention of nosocomial influenza infection with amantadine. *Am. J. Epidemiol.*, **97**, 276–282

Oker-Blom, N. and Andersen, L. (1966). Effect of 1-adamantanamine hydrochloride on Rous sarcoma virus *in vitro* and *in vivo*. *Eur. J. Cancer*, **2**, 9–12

Oker-Blom, N., Hovi, T., Leinikki, P., Palosuo, T., Pettersson, R. and Suni, J. (1970). Protection of man from natural infection with influenza A2 Hong Kong virus by amantadine: a controlled field trial. *Br. Med. J.*, **3**, 676–678

Oki, T. and Heidelberger, C. (1971). Fluorinated pyrimidines. XXXIX. Effects of 5-trifluoromethyl-2′-deoxyuridine on the replication of vaccinia viral messenger ribonucleic acid and proteins. *Mol. Pharmacol.*, **7**, 653–662

Omelchenko, T. N., Čiumpov, F., Nadtochey, G. A., Avakyan, A. A. and Altshtein, A. D. (1969). Effects of inhibitors of DNA synthesis on haemagglutinin and infectious SV15 virus formation in green monkey kidney cells. *Acta Virol.*, **13**, 461–468

O'Neill, F. J. and Rapp, F. (1971). Synergistic effect of herpes simplex virus and cytosine arabinoside on human chromosomes. *J. Virol.*, **7**, 692–695

Oxford, J. S., Logan, I. S. and Potter, C. W. (1970). *In vivo* selection of an influenza A2 strain resistant to amantadine. *Nature*, **226**, 82–83

Oxford, J. S. and Perrin, D. D. (1974). Inhibition of the particle-associated RNA-dependent RNA polymerase activity of influenza virus by chelating agents. *J. Gen. Virol.*, **23**, 59–71

Oxford, J. S. and Potter, C. W. (1973). Aminoadamantane-resistant strains of influenza A2 virus. *J. Hyg., Camb.*, **71**, 227–235

Oxford, J. S. and Schild, G. C. (1965). *In vitro* inhibition of rubella virus by 1-adamantanamine hydrochloride. *Arch. Ges. Virusforsch.*, **17**, 313–329

Oxford, J. S. and Schild, G. C. (1967 a). The evaluation of antiviral compounds for rubella virus using organ cultures. *Arch. Ges. Virusforsch.*, **22**, 349–356

Oxford, J. S. and Schild, G. C. (1967 b). Inhibition of the growth of influenza and rubella viruses by amines and ammonium salts. *Br. J. Exp. Path.*, **48**, 235–243

Oyanagi, J., ter Meulen, V., Katz, M. and Koprowski, H. (1971). Comparison of subacute sclerosing panencephalitis and measles virus: an electron microscope study. *J. Virol.*, **7**, 176–187

Padgett, B. L., Walker, D. L., ZuRhein, G. M., Eckroade, R. J. and Dessel, B. H. (1971). Cultivation of papova-like virus from human brain with progressive multifocal leukoencephalopathy. *Lancet*, **i**, 1257–1260

Palm, P. E., Nick, M. S., Funkhauser, J. J. and Kensler, C. J. (1967). Toxicological studies on repeated intravenous dosages of 5-trifluoromethyl-2'-deoxyuridine (FT_3DR) in dogs and monkeys. *Toxicol. Appl. Pharmacol.*, **10**, 406–407

Papac, R. J., Calabresi, P., Hollingsworth, J. W. and Welch, A. D. (1965). Effects of 1-β-D-arabinofuranosylcytosine hydrochloride on regenerating bone marrow. *Cancer Res.*, **25**, 1459–1462

Patterson, A. and Jones, B. R. (1967). The management of ocular herpes. *Trans. Ophthal. Soc.*, **87**, 59–84

Pavan-Langston, D. and Dohlman, C. H. (1972). A double-blind clinical study of adenine arabinoside therapy of viral keratoconjunctivitis. *Am. J. Ophthal.*, **74**, 81–88

Pavan-Langston, D., Dohlman, C. H., Geary, P. and Sulzewski, D. (1973). Intraocular penetration of ara A and IDU—therapeutic implications in clinical herpetic uveitis. *Trans. Am. Acad. Ophthal. Otolaryngol.*, **77**, 455–466

Pavan-Langston, D. and McCulley, J. P. (1973). Herpes zoster dendritic keratitis. *Arch. Ophthal.*, **89**, 25–29

Payne, F. E., Baublis, J. V. and Itabashi, H. H. (1969). Isolation of measles virus from cell cultures of brain from a patient with subacute sclerosing panencephalitis. *N. Engl. J. Med.*, **281**, 585–589

Perkins, E. S., Wood, R. M., Sears, M. L., Prusoff, W. H. and Welch, A. D. (1962). Anti-viral activities of several iodinated pyrimidine deoxyribonucleosides. *Nature*, **194**, 985–986

Person, D. A., Sheridan, P. J. and Herrmann, E. C. (1970). Sensitivity of types 1 and 2 herpes simplex virus to 5-iodo-2'-deoxyuridine and 9-β-D-arabinofuranosyladenine. *Infect. Immunity*, **2**, 815–820

Pfau, C. J., Trowbridge, R. S., Welsh, R. M., Staneck, L. D. and O'Connell, C. M. (1972). Arenaviruses: inhibition by amantadine hydrochloride. *J. Gen. Virol.*, **14**, 209–211

Pinkel, D. (1958). The use of body surface area as a criterion of drug dosage in cancer chemotherapy. *Cancer Res.*, **18**, 853–856

Plotkin, S. A. and Stetler, H. (1969). Treatment of congenital cytomegalic inclusion disease with antiviral agents. *Antimicrob. Ag. Chemother.*, 372–379

Plowright, W., Brown, F. and Parker, J. (1966). Evidence for the type of nucleic acid in African swine fever virus. *Arch. Ges. Virusforsch.*, **19**, 288–304

Polatnick, J. (1965). Effect of chemical agents on foot-and-mouth disease virus production in cell cultures. *Am. J. Vet. Res.*, **26**, 1051–1055

Prager, D., Bruder, M. and Sawitsky, A. (1971). Disseminated varicella in a patient with acute myelogenous leukemia: treatment with cytosine arabinoside. *J. Pediatr.*, **78**, 321–323

Prokhorova, A. M. and Solov'ev, V. N. (1967). Organotropic properties of 1-adamantanamine by different routes of administration. *Pharmakol. Toksikol.*, **30**, 203–206

Prusoff, W. H. (1963). A review of some aspects of 5-iododeoxyuridine and azauridine. *Cancer Res.*, **23**, 1246–1259

Prusoff, W. H., Bakhle, Y. S. and Sekely, L. (1965). Cellular and antiviral effects of halogenated deoxyribonucleosides. *Ann. N.Y. Acad. Sci.*, **130**, 135–150

Prusoff, W. H. and Goz, B. (1973). Chemotherapy—Molecular aspects. In: A. S. Kaplan (ed.) *The Herpesviruses*, 641–663. (New York and London: Academic Press Inc.)

Prusoff, W. H., Jaffe, J. J. and Günther, H. (1960). Studies in the mouse of the pharmacology of 5'-iododeoxyuridine. *Biochem. Pharmacol.*, **3**, 110–121

Quilligan, J. J., Hirayama, M. and Bernstein, H. D. (1966). The suppression of A2 influenza in children by the chemoprophylactic use of amantadine. *J. Pediatr.*, **69**, 572–575

Rada, B., Blaškovič, D., Šorm, F. and Škoda, J. (1960). The inhibitory effect of 6-azauracil riboside on the multiplication of vaccinia virus. *Experientia*, **16**, 487

Rao, A. R., Jacobs, E. S., Kamalakshi, S., Bradbury and Swamy, A. (1969 a). Chemoprophylaxis and chemotherapy in variola major. Part I. An assessment of CG 662 and Marboran in prophylaxis of contacts of variola major. *Indian J. Med. Res.*, **57**, 477–483

Rao, A. R., Jacobs, E. S., Kamalakshi, S., Bradbury and Swamy, A. (1969 b). Chemoprophylaxis and chemotherapy in variola major. Part II. Therapeutic assessment of CG 662 and Marboran in treatment of variola major in man. *Indian J. Med. Res.*, **57**, 484–494

Rao, R. N. and Natarajan, A. T. (1965). Effect of 9-β-D-arabinofuranosyladenine on *Vicia faba* chromosomes. *Cancer Res.*, **25**, 1764–1769

Rapp, F. (1964). Inhibition by metabolic analogues of plaque formation by herpes zoster and herpes simplex viruses. *J. Immunol.*, **93**, 643–648

Rapp, F., Melnick, J. L. and Kitahara, T. (1965). Tumor and virus antigens of simian virus 40: differential inhibition of synthesis by cytosine arabinoside. *Science*, **147**, 625–627

Ravin, L. F., Simpson, C. A., Zappala, A. F. and Gulesich, J. J. (1964). Hydrolysis of idoxuridine. *J. Pharm. Sci.*, **53**, 1064–1066

Rawls, W. E., Cohen, R. A. and Herrmann, E. C. (1964). Inhibition of varicella virus by 5-iodo-2'-deoxyuridine. *Proc. Soc. Exp. Biol. Med.*, **115**, 123–127

Rawls, W. E., Melnick, J. L., Olson, G. B., Dent, P. B. and Good, R. A. (1967). Effect of amantadine hydochloride on the response of human lymphocytes to phytohaemagglutinins. *Science*, **158**, 506–507

Renis, H. E. and Johnson, H. G. (1962). Inhibition of plaque formation of vaccinia virus by cytosine arabinoside hydrochloride. *Bact. Proc.*, **140**

Report of the Committee of Enquiry into the smallpox outbreak in London in March and April 1970. H.M. Stationery Office, London, 1974

Reyes, P. and Heidelberger, C. (1965). Fluorinated pyrimidines. XXVI. Mammalian thymidylate synthetase: its mechanism of action and inhibition by fluorinated nucleosides. *Mol. Pharmacol.*, **1**, 14–30

Reynolds, D. W., Stagno, S., Hosty, T. S., Tiller, M. and Alford, C. A. (1973). Maternal cytomegalovirus excretion and perinatal infection. *N. Engl. J. Med.*, **289**, 1–5

Rhim, J. S., Lane, W. T. and Huebner, R. J. (1972). Amantadine hydrochloride: inhibitory effect on murine sarcoma virus infection in cell cultures. *Proc. Soc. Exp. Biol. Med.*, **139**, 1258–1260

Ribeiro do Valle, L. A., Raposo de Melo, P., de Salles Gomes, L. F. and Morato Proença, L. (1965). Methisazone in prevention of variola minor among contacts. *Lancet*, **ii**, 976–978

Rice, N. S. L. and Jones, B. R. (1973). Problems of corneal grafting in herpetic keratitis. 220–239. *Corneal Graft Failure*: Ciba Foundation Symposium. (Amsterdam: Elsevier)

Rifkind, D., Starzl, T. E., Marchions, T. L., Weddell, W. R., Rowlands, D. T. and Hill, R. B. (1964). Transplantation pneumonia. *J. Am. Med. Ass.*, **189**, 808–812

Roberts, D. and Loehr, E. V. (1972). Depression of thymidylate synthetase activity in response to cytosine arabinoside. *Cancer Res.*, **32**, 1160–1169

Roby, R., Teskey, C. and Houlihan, R. B. (1965). Inhibition of plaque formation by myxoma and fibroma viruses with pyrimidine analogues. *Proc. Soc. Exp. Biol. Med.*, **120**, 496–500

Rogers, W. I., Hartman, A. C., Palm, P. E., Okstein, C. and Kensler, C. J. (1969). The fate of 5-trifluoromethyl-2'-deoxyuridine in monkeys, dogs, mice and tumor-bearing mice. *Cancer Res.*, **29**, 953–961

Rogers, W. I. and Wilson, J. A. (1969). Determination of labile trifluoromethyl compounds with a fluoride-ion electrode: differential analysis of 5-trifluoromethyluracil and 5-trifluoromethyl-2'-deoxyuridine. *Analyt. Biochem.*, **32**, 31–37

Rowe, W. P., Hartley, J. W., Waterman, S., Turner, H. C. and Huebner, R. J. (1956). Cytopathogenic agent resembling human salivary gland virus recovered from tissue cultures of human adenoids. *Proc. Soc. Exp. Biol. Med.*, **92**, 418–424

Sabin, A. B. (1967). Amantadine hydrochloride. Analysis of data related to its proposed use for prevention of A2 influenza virus disease in human beings. *J. Am. Med. Ass.*, **200**, 943–950

Sabin, A. B. and Wright, A. M. (1934). Acute ascending myelitis following a monkey bite, with the isolation of a virus capable of reproducing the disease. *J. Exp. Med.*, **59**, 115–136

Scatton, B., Cheramy, A., Besson, M. J. and Glowinski, J. (1970). Increased synthesis and release of dopamine in the striatum of the rat after amantadine treatment. *Eur. J. Pharmacol.*, **13**, 131–133

Schabel, F. M. (1968). The antiviral activity of 9-β-D-arabinofuranosyladenine (ARA-A). *Chemotherapy*, **13**, 321–338

Schiff, G., Bloomfield, S. and Gaffney, T. (1966). The effects of amantadine HCl on experimental human influenza. *Clin. Res.*, **14,** 343

Schild, G. C. and Sutton, R. N. P. (1965). Inhibition of influenza viruses *in vitro* and *in vivo* by 1-adamantanamine hydrochloride. *Br. J. Exp. Pathol.*, **46,** 263–273

Schimpff, S. C., Fortner, C. L., Greene, W. H. and Wiernik, P. H. (1974). Cytosine arabinoside for localized herpes zoster in patients with cancer: failure in a controlled trial. *J. Infect. Dis.*, **130,** 673–676

Schnebli, H. P., Hill, D. L. and Bennett, L. L. (1967). Purification and properties of adenosine kinase from human tumor cells of type HEp No. 2. *J. Biol. Chem.*, **242,** 1997–2004

Schrecker, A. W. and Urshel, M. J. (1968). Metabolism of 1-β-D-arabino-furanosylcytosine in leukemia L1210: studies with intact cells. *Cancer Res.*, **28,** 793–801

Schwab, R. S., England, A. C., Poskanzer, D. C. and Young, R. R. (1969). Amantadine in the treatment of Parkinson's disease. *J. Am. Med. Ass.*, **208,** 1168–1170

Sheffield, F. W., Bauer, D. J. and Stephenson, S. M. (1960): The protection of tissue cultures by isatin β-thiosemicarbazone from the cytopathic effects of certain pox viruses. *Br. J. Exp. Pathol.*, **41,** 638–647

Shen, T. Y., McPherson, J. F. and Linn, B. O. (1966). Nucleosides. III. Studies on 5-methylamino-2'-deoxyuridine as a specific antiherpes agent. *J. Med. Chem.*, **9,** 366–369

Shipman, C., Novack, J. M. and Drach, J. C. (1973). Effect of 9-β-D-arabino-furanosyladenine (ara-A) and 9-β-D-arabinofuranosylhypoxanthine (ara-Hx) on growth rate and nucleic acid synthesis in B-mix K-44/6 cells. *Fed. Proc.*, Abstr., 2737, 699

Sidwell, R. W., Arnett, G., Dixon, G. J. and Schabel, F. M. (1969). Purine analogs as potential anticytomegalovirus agents. *Proc. Soc. Exp. Biol. Med.*, **131,** 1223–1230

Sidwell, R. W., Arnett, G. and Schabel, F. M. (1970). Effects of 9-β-D-arabino-furanosyladenine on myxoma and pseudorabies viruses. *Prog. Antimicrob. Anticancer Chemother.*, **2,** 44–48

Sidwell, R. W., Arnett, G. and Schabel, F. M. (1972). *In vitro* effect of a variety of biologically active compounds on human cytomegalovirus. *Chemotherapy*, **17,** 259–282

Silagi, S. (1965). Metabolism of 1-β-D-arabinofuranosylcytosine in L cells. *Cancer Res.*, **25,** 1446–1453

Silverman, L. and Rubinstein, L. J. (1965). Electron microscopic observations on a case of progressive multifocal leukoencephalopathy. *Acta Neuropathol.*, **5,** 215–224

Simpson, C. A. and Zappala, A. F. (1964). Procedure for assay and stability determination of idoxuridine. *J. Pharm. Sci.*, **53,** 1201–1204

Skalko, R. G. and Packard, D. S. (1973). The teratogenic response of the mouse embryo to 5-iododeoxyuridine. *Experientia*, **29,** 198–200

Smith, C. G., Buskirk, H. H. and Lummis, W. L. (1967). Nucleic acids II: cytotoxicity studies with nucleotides and dinucleotide phosphates containing *ara*-cytidine. *J. Med. Chem.*, **10,** 774–776

Smith, C. G., Lummis, W. L. and Grady, J. E. (1959). An improved tissue culture assay. I. Methodology and cytotoxicity of anti-tumor agents. *Cancer Res.*, **19**, 843–846

Smith, K. O. and Dukes, C. D. (1964). Effects of 5-iodo-2-desoxyuridine (IDU) on herpesvirus synthesis and survival in infected cells. *J. Immunol.*, **92**, 550–554

Smorodintsev, A. A., Karpuhin, G. I., Zlydnikov, D. M., Malyševa, A. M., Švecova, E. C., Burov, S. A., Hramcova, L. M., Romanov, J. A., Taros, L. J., Ivannikov, J. G. and Novoselov, S. D. (1970 a). The prophylactic effectiveness of amantadine hydrochloride in an epidemic of Hong Kong influenza in Leningrad in 1969. *Bull. Wld. Hlth. Org.*, **42**, 865–872

Smorodintsev, A. A., Zlydnikov, D. M., Kiseleva, A. M., Romanov, Yu.A., Kasantsev, A. P. and Rumovsky, V. I. (1970 b). Evaluation of amantadine in artificially induced A2 and B influenza. *J. Am. Med. Ass.*, **213**, 1448–1454

Smorodintsev, A. A., Zlydnikov, D. M., Romanov, Yu.A. and Rumovsky, V. I. (1972). Effectiveness of amantadine hydrochloride (midantane) in the prevention of artificially induced influenza infection. *Vopr. Virusol.*, **17**, 152–156

Stephenson, J. A., Artenstein, M. S., Parkman, P. D., Buescher, E. L. and Druzd, A. D. (1965). Effect of amantadine hydrochloride on rubella virus infection in the rhesus monkey. *Antimicrob. Ag. Chemother.*, 548–552

Stern, H. and Elek, S. D. (1965). The incidence of infection with cytomegalovirus in a normal population. *J. Hyg., Camb.*, **63**, 79–87

Stern, H., Elek, S. D., Millar, D. M. and Anderson, H. F. (1959). Herpetic whitlow. A form of cross-infection in hospitals. *Lancet*, **ii**, 871–874

Stevens, D. A., Jordan, G. W., Waddell, T. F. and Merigan, T. C. (1973). Adverse effect of cytosine arabinoside on disseminated zoster in a controlled trial. *N. Engl. J. Med.*, **289**, 873–878

Stevens, J. G. and Cook, M. L. (1973 a). Latent herpes simplex virus in spinal ganglia of mice. *Science*, **173**, 843–845

Stevens, J. G. and Cook, M. L. (1973 b). Latent infections induced by herpes simplex virus. *Cancer Res.*, **33**, 1399–1401

St. Jeor, S. and Rapp, F. (1973 a). Cytomegalovirus replication in cells pretreated with 5-iodo-2'-deoxyuridine. *J. Virol.*, **11**, 986–990

St. Jeor, S. and Rapp, F. (1973 b). Cytomegalovirus: conversion of nonpermissive cells to a permissive state for virus replication. *Science*, **181**, 1060–1061

Strömberg, U., Svensson, T. H. and Waldeck, B. (1970). On the mode of action of amantadine. *J. Pharm. Pharmacol.*, **22**, 959–962

Sugar, J., Varnell, E., Centifanto, Y. and Kaufman, H. E. (1973). Trifluorothymidine treatment of herpetic iritis in rabbits and ocular penetration. *Invest. Ophthal.*, **12**, 532–534

Svensson, T. H. and Strömberg, U. (1970). Potentiation by amantadine hydrochloride of L-dopa-induced effects in mice. *J. Pharm. Pharmacol.*, **22**, 639–640

Szybalski, W., Cohn, N. K. and Heidelberger, C. (1963). Effects of 5-trifluoromethyl-2'-deoxyuridine (F_3TdR) on the biochemical and radiobiological properties of human cells. *Fed. Proc.*, **22**, 532

Talley, R. W., O'Bryan, R. M., Tucker, W. G. and Loo, R. V. (1967). Clinical pharmacology and human antitumor activity of cytosine arabinoside. *Cancer*, **20**, 809–816

Talley, R. W. and Vaitkevicius, V. K. (1963). Megaloblastosis produced by a cytosine antagonist 1-β-D-arabinofuranosylcytosine. *Blood*, **21**, 352–361

Talley, R. W., Vaitkevicius, V. K., Reed, M. L. and Brennan, M. J. (1962). Cytosine arabinoside; human pharmacology and toxicity. *Proc. Am. Ass. Cancer Res.*, **3**, 366

Tarsy, D., Holden, E. M., Segarra, J. M., Calabresi, P. and Feldman, R. G. (1973). 5-Iodo-2'-deoxyuridine (IUDR: NSC-39661) given intraventricularly in the treatment of progressive multifocal leukoencephalopathy. *Cancer Chemother. Rep.*, Pt. 1, **57**, 73–78

Ter Meulen, V., Iwasaki, Y., Kaprowski, H., Käckell, Y. M. and Müller, D. (1972 a). Fusion of cultured multiple-sclerosis brain cells with indicator cells: presence of nucleocapsids and virions and isolation of parainfluenza-type virus. *Lancet*, **ii**, 1–5

Ter Meulen, V., Leonard, L. L., Lennette, E. H., Katz, M. and Kaprowski, H. (1972 b). The effect of 6-azauridine upon subacute sclerosing panencephalitis virus in tissue culture. *Proc. Soc. exp. Biol., Med.*, **140**, 1111–1115

Thiel, von N., L'Age-Stehr, J. and Wacker, A. (1967). Untersuchungen über den Einfluss von Hemmstoffen der Proteinsynthese auf die Bildung von Antikörpern. *Hoppe-Seyler's Z. Physiol. Chem.*, **348**, 1407–1414

Thompson, R. L., Minton, S. A., Officer, J. E. and Hitchings, G. H. (1953). Effect of heterocyclic and other thiosemicarbazones on vaccinia infection in the mouse. *J. Immunol.*, **70**, 229–234

Togo, Y., Hornick, R. B. and Dawkins, A. T. (1968). Studies on induced influenza in man. I. Double-blind studies designed to assess prophylactic efficacy of amantadine hydrochloride against A2/Rockville/1/65 strain. *J. Am. Med. Ass.*, **203**, 1089–1094

Togo, Y., Hornick, R. B., Felitti, V. J., Kaufman, M. L., Dawkins, A. T., Kilpe, V. E. and Claghorn, J. L. (1970). Evaluation of therapeutic efficacy of amantadine in patients with naturally occurring A2 influenza. *J. Am. Med. Ass.*, **211**, 1149–1156

Trainin, N., Kaye, A. M. and Berenblum. I. (1964). Influence of mutagens on the initiation of skin carcinogenesis. *Biochem. Pharmacol.*, **13**, 263–267

Tsubouci, S., Goromaru, T. and Iguchi, S. (1970). Microanalysis of amantadine. *Bunseki Kagaku*, **19**, 95–98 [from *Chem. Abstr.*, **73**, 12723h, 1970]

Turner, W., Bauer, D. J. and Nimmo-Smith, R. H. (1962). Eczema vaccinatum treated with N-methylisatin β-thiosemicarbazone. *Br. Med. J.*, **1**, 1317–1319

Tyrrell, D. A. J., Bynoe, M. L. and Hoorn, B. (1965). Studies on the antiviral activity of adamantanamine. *Br. J. Exp. Pathol.*, **46**, 370–375

Uchiyama, M. and Shibuya, M. (1969). Distribution and excretion of ^{3}H-amantadine HCl. *Chem. Pharm. Bull.*, **17**, 841–843

Umeda, M. and Heidelberger, C. (1969). Fluorinated pyrimidines. XXXI. Mechanisms of inhibition of vaccinia virus replication in HeLa cells by pyrimidine nucleosides. *Proc. Soc. Exp. Biol. Med.*, **130**, 24–29

Underwood, G. E. (1962). Activity of 1-β-D-arabinofuranosylcytosine hydro-chloride against herpes simplex keratitis. *Proc. Soc. Exp. Biol. Med.*, **111**, 660–664

Underwood, G. E., Elliott, G. A. and Buthala, D. A. (1965). Herpes keratitis in rabbits: pathogenesis and effect of antiviral nucleosides. *Ann. N.Y. Acad. Sci.*, **130**, Art. 1, 151–167

Váczi, L. and Gönczöl, E. (1973). The effect of cytosine-arabinoside on the multiplication of cytomegalovirus and on the formation of virus-induced intracellular antigens. *Acta Virol.*, **17**, 189–195

Van Rooyen, C. E., Casey, J., Lee, S. H. S., Faulkner, R. and Dincsoy, H. P. (1967). Vaccinia gangrenosa and 1-methylisatin 3-thiosemicarbazone (methisazone). *Can. Med. Ass. J.*, **97**, 160–165

Vernier, V. G., Harmon, J. B., Stump, J. M., Lynes, T. E., Marvell, J. P. and Smith, D. H. (1969). The toxicological and pharmacological properties of amantadine hydrochloride. *Toxicol. Appl. Pharmacol.*, **15**, 642–665

Wallbank, A. M., Matter, R. E. and Klinikowski, N. G. (1966). 1-Adamantana-mine hydrochloride: inhibition of Rous and Esh sarcoma viruses in cell culture. *Science*, **152**, 1760–1761

Waltuch, G. and Sachs, F. (1968). Herpes zoster in a patient with Hodgkin's disease. *Arch. Intern. Med.*, **121**, 458–462

Watson, D. H. (1973). Replication of the viruses—morphological aspects. In: A. S. Kaplan (ed.) *The Herpesviruses*, 133–161. (London: Academic Press)

Webb, D. R., Bourne, H. R. and Levinson, W. (1974). A new phosphodiesterase inhibitor in human lymphocytes: *N*-methyl-isatin-β-thiosemicarbazone. *Biochem. Pharmacol.*, **23**, 1663–1667

Webb, J. F., Marks, R. and Reed, T. A. G. (1965). Eczema vaccinatum treated with *N*-methylisatin β-thiosemicarbazone. *Br. J. Dermatol.*, **77**, 596–598

Weiner, L. P., Herndon, R. M., Narayan, O., Johnson, O. T., Shah, K., Rubinstein, L. J., Preziosi, T. J. and Conley, F. K. (1972). Isolation of virus related to SV40 from patients with progressive multifocal leukoencephalo-pathy. *N. Engl. J. Med.*, **286**, 385–390

Welch, A. D., Jaffe, J. J., Cardoso, S. S., Finch, S. C., Calabresi, P., Liebow, A. A. and Prusoff, W. H. (1960). Studies on the pharmacology of iododeoxyuridine in animals and man. *Proc. Am. Ass. Cancer Res.*, **3**, 161

Weller, T. H. and Coons, A. H. (1954). Fluorescent antibody studies with agents of varicella and herpes zoster propagated *in vitro*. *Proc. Soc. Exp. Biol. Med.*, **86**, 789–794

Wellings, P. C., Awdry, P. N., Bors, F. H., Jones, B. R., Brown, D. C. and Kaufman, H. E. (1972). Clinical evaluation of trifluorothymidine in the treatment of herpes simplex corneal ulcers. *Am. J. Ophthal.*, **73**, 932–942

Welsh, R. M., Trowbridge, R. S., Kowalski, J. B., O'Connell, C. M. and Pfau, C. J. (1971). Amantadine hydrochloride inhibition of early and late stages of lymphocytic choriomeningitis virus-cell interactions. *Virology*, **45**, 679–686

Wendel, H. A., Snyder, M. T. and Pell, S. (1966). Trial of amantadine in epidemic influenza. *Clin. Pharm. Ther.*, **7**, 38–43

Wesemann, W. and Zilliken, F. (1967). Adamantanamines and their derivatives as sensitizing agents for 5-hydroxytryptamine-induced contractions of smooth muscle. *J. Pharm. Pharmacol.*, **19**, 203–207

White, C. M. (1963). Vaccinia gangrenosa due to hypogammaglobulinaemia. *Lancet*, **i**, 969–971

Wilkerson, S., Finley, S. C., Finley, W. H. and Ch'ien, L. T. (1973). Chromosome breakage in patients receiving ara-A. *Clin. Res.*, **21**, 52

Willcox, R. R. (1968). Necrotic cervicitis due to primary infection with the virus of herpes simplex. *Br. Med. J.*, **1**, 610–612

Wingfield, W. L., Pollack, D. and Grunert, R. R. (1969). Therapeutic efficacy of amantadine HCl and rimantadine HCl in naturally occurring influenza A2 respiratory illness in man. *N. Engl. J. Med.*, **281**, 579–584

York, J. L. and LePage, G. A. (1966). A proposed mechanism for the action of 9-β-D-arabinofuranosyladenine as an inhibitor of the growth of some ascites cells. *Can. J. Biochem.*, **44**, 19–26

Yoshikura, H. (1968). Requirement of cellular DNA synthesis for the growth of Friend leukemia virus. *Exp. Cell Res.*, **52**, 445–450

Zaky, D. A., Betts, R. F., Douglas, R. G., Bengali, K. and Neil, G. L. (1975). Varicella-zoster virus and subcutaneous cytarabine: correlation of *in vitro* sensitivities to blood levels. *Antimicrob. Ag. Chemother.*, **7**, 229–232

Zischka-Konorsa, W., Jellinger, K. and Hohenegger, M. (1965). Zur Pathogenese von Herpesvirus-Erkrankungen mit besonderer Berücksichtigung der nekrotisierenden Herpes simplex-Encephalitis. *Acta Neuropathol.*, **5**, 252–274

ZuRhein, G. M. and Chou, S.-M. (1965). Particles resembling papova viruses in human cerebral demyelinating disease. *Science*, **148**, 1477–1479

Index